GENOMIC BIOMARKERS FOR PHARMACEUTICAL DEVELOPMENT

GENOMIC BIOMARKERS FOR PHARMACEUTICAL DEVELOPMENT

ADVANCING PERSONALIZED HEALTH CARE

Edited by

YIHONG YAO, PhD
MedImmune, LLC
Gaithersburg, Maryland

BAHIJA JALLAL, PhD
MedImmune, LLC
Gaithersburg, Maryland

KOUSTUBH RANADE, PhD
MedImmune, LLC
Gaithersburg, Maryland

AMSTERDAM • BOSTON • HEIDELBERG • LONDON • NEW YORK • OXFORD
PARIS • SAN DIEGO • SAN FRANCISCO • SINGAPORE • SYDNEY • TOKYO
Academic Press is an imprint of Elsevier

Academic Press is an imprint of Elsevier
525 B Street, Suite 1900, San Diego, CA 92101-4495, USA
32 Jamestown Road, London NW1 7BY, UK
225 Wyman Street, Waltham, MA 02451, USA

The content for this book was provided by Yihong Yao, Bahija Jallal and Koustubh Ranade. Drs. Yao,
Jallal and Ranade are employees of MedImmune, LLC. The views and opinions in this book are their
personal views and opinions and are not endorsed by MedImmune.

Notice
No responsibility is assumed by the publisher for any injury and/or damage to persons, or property
as a matter of products liability, negligence or otherwise, or from any use or, operation of any
methods, products, instructions or ideas contained in the material herein. Because of rapid advances
in the medical sciences, in particular, independent verification of diagnoses and drug dosages should
be made.

British Library Cataloging-in-Publication Data
A catalog record for this book is available from the British Library

Library of Congress Cataloging-in-Publication Data
A catalog record for this book is available from the Library of Congress

ISBN: 978-0-12-397336-8

For information on all Academic Press publications
visit our website at elsevierdirect.com

Typeset by MPS Limited, Chennai, India
www.adi-mps.com

Printed and bound in the United States of America

14 15 16 17   10 9 8 7 6 5 4 3 2 1

Working together
to grow libraries in
developing countries

www.elsevier.com • www.bookaid.org

Contents

Preface

A number of books have been published lately on genomics and personalized medicine. We felt, however, that none was comprehensive and reflected adequately our thinking and lived experience in applying genomics to drug development and personalized healthcare. Furthermore, published literature was also relatively silent on the translational science approaches to identify the right therapeutic for the right patient that we believe will be needed to improve the probability of success of clinical trials, particularly in the early stages of drug development.

We made an effort to address these gaps in this book by describing genomic strategies to explore basic questions in drug development: What kind of patient might most benefit from a new therapeutic? Which patient might experience adverse effects? How does one identify such patients? How does one determine if a new therapeutic is having the desired effect even before clinical symptoms improve? The most current thinking on these questions from industry, academia and the government is assembled in this book.

The book begins with a broad discussion of how to employ genomics to develop translational approaches for early stage clinical trials and to understand adverse side effects of therapeutics. We include case studies to illustrate key points. The ensuing chapters describe in detail how strategies outlined in the first chapter may be reduced to practice in different therapeutic areas.

Given the vast investment in the war on cancer over the past 40 years, and the breathtaking progress in understanding the complex landscape of the cancer genome, it should come as no surprise that personalized healthcare based on genomics is most mature for this therapeutic area. As the chapter on cancer makes clear, however, even in this area, significant hurdles need to be overcome before individualized therapies come into routine clinical practice.

The chapter on autoimmune diseases describes early steps in developing potential personalized healthcare for rheumatoid arthritis and systemic lupus erythematosus, and the use of genomic biomarkers to determine if a therapeutic is having the intended effect. Chapter 4 is a tour de force of how to apply genomics to tease apart the complexity of asthma and turn that knowledge into a targeted therapeutic. The nascent fields of microRNAs – small RNA molecules that regulate the expression of a large number of genes – and toxicogenomics are the subjects of the next two chapters. The book ends with a discussion of the regulatory process for obtaining approval for diagnostic tests that are necessary for personalized healthcare. As our understanding of cardiovascular diseases, mental illness and rare genetic disorders deepens, we look forward to discussing targeted therapeutics for these diseases in a future editions of this book.

The book is intended for anyone interested in understanding current and emerging practices in translational science as they relate to genomics and drug development. It is appropriate for advanced graduate students, clinicians and scientists working in the bio-pharmaceutical industry.

We are very grateful to the co-authors for their efforts and thank our families for tolerating the many evenings and weekends that were devoted to this book.

Yihong Yao
Bahija Jallal
Koustubh Ranade

About the Editors

Dr. Yihong Yao is director and head of Pharmacogenomics and Bioinformatics group at MedImmune, LLC. The focus of Dr. Yao's group is the utilization of cutting edge genomic and genetic approaches to developing pharmacodynamic and predictive diagnostic markers to understand disease linkage and to identify the right patients that might respond (or not respond) to therapeutic interventions. The other areas of interest in Dr. Yao's group include: unveiling potential key drivers in cancer, and in respiratory and inflammatory diseases, and understanding the role of miRNAs in disease pathogenesis.

Dr. Yao received a Bachelor's degree in Biochemistry from Fu Dan University in 1988. He received a Master's degree in Bioinformatics from Boston University. He completed his doctorate in Biochemistry and Biophysics from the University of Kansas in Lawrence, Kansas. He conducted his postdoctoral research at Johns Hopkins Medical School in Baltimore, Maryland.

Dr. Yao has authored over 50 peer-reviewed publications, edited a book and has over 15 patents.

Dr. Bahija Jallal is Executive Vice President of AstraZeneca and head of MedImmune, a global biologics research and development organization with locations in Gaithersburg, California and Cambridge, UK. She is a member of the senior executive team at AstraZeneca reporting to the CEO. Dr. Jallal joined MedImmune in March 2006. She has guided the MedImmune R&D organization through the unprecedented growth and expansion of its biologics pipeline from 40 drugs to more than 120 today. She inspires creativity, out-of-the-box thinking and a dedication to scientific excellence in the more than 2,500 employees at MedImmune.

Dr. Jallal received a Master's degree in biology from the Universite de Paris VII in France, and her doctorate in physiology from the University of Pierre & Marie Curie in Paris. She conducted her postdoctoral research at the Max-Planck Institute of Biochemistry in Martinsried, Germany.

Dr. Jallal has authored over 70 peer-reviewed publications and has over 15 patents. She is a member of the American Association of Cancer Research, the American Association of Science and the Pharmacogenomics Working Group. She serves as a member of the Board of Directors for the Association of Women in Science and is an advisory member of the Healthcare Business Women's Association. She was named one of FierceBiotech's 2011 Women in Biotech and in 2012, she received Washington Business Journal's Women Who Mean Business Award.

Dr. Koustubh Ranade is Fellow in Translational Sciences at MedImmune, where he is responsible for leading translational and personalized healthcare strategies for multiple drug development programs across therapeutic areas. Prior to MedImmune, Koustubh was Associate Group Medical Director of Clinical Diagnostics at Genentech and Director of Human Genetics at Bristol-Myers Squibb. Dr Ranade has published extensively on human genetics and pharmacogenomics and is inventor on four issued patents. He received his PhD in Molecular Genetics from the University of Massachusetts Medical School, and did his postdoctoral training at the National Human Genome Research Institute and Stanford University.

List of Contributors

Alex Adai Asuragen, Inc., Austin, Texas

Bernard Andruss Asuragen, Inc., Austin, Texas

Joseph R. Arron Genentech, Inc., South San Francisco, California

Jurgen Borlak Center of Pharmacology and Toxicology, Hannover Medical School, Hannover, Germany

Philip Brohawn MedImmune, LLC, Gaithersburg, Maryland

Nicholas C. Dracopoli Janssen Research and Development, LLC Radnor, Pennsylvania

Jeffrey M. Harris Genentech, Inc., South San Francisco, California

Brandon W. Higgs MedImmune, LLC, Gaithersburg, Maryland

Bahija Jallal MedImmune, LLC, Gaithersburg, Maryland

Gary Latham Asuragen, Inc., Austin, Texas

Elizabeth Mambo Asuragen, Inc., Austin, Texas

Ruth March AstraZeneca Pharmaceuticals, Research and Development Genetics Department, Mereside, Macclesfield, Cheshire, United Kingdom

Koustubh Ranade MedImmune, LLC, Gaithersburg, Maryland

Lorin Roskos MedImmune, LLC, Gaithersburg, Maryland

Annette Schlageter Asuragen, Inc., Austin, Texas

Katie Streicher MedImmune, LLC, Gaithersburg, Maryland

Anna E. Szafranska-Schwarzbach Asuragen, Inc., Austin, Texas

Weida Tong Division of Bioinformatics and Biostatistics, National Center for Toxicological Research, US Food and Drug Administration, Jefferson, Arkansas

Cornelis L. Verweij Department of Pathology, VU University Medical Center, Amsterdam, The Netherlands

Yuping Wang Division of Bioinformatics and Biostatistics, National Center for Toxicological Research, US Food and Drug Administration, Jefferson, Arkansas

Yihong Yao MedImmune, LLC, Gaithersburg, Maryland

Application of Translational Science to Clinical Development

Koustubh Ranade[1], Brandon W. Higgs[1], Ruth March[2], Lorin Roskos[1], Bahija Jallal[1], Yihong Yao[1]

[1]MedImmune, LLC, Gaithersburg, Maryland

[2]AstraZeneca Pharmaceuticals, Research and Development Genetics Department, Mereside, Macclesfield, Cheshire, United Kingdom

1.1 INTRODUCTION

The pharmaceutical industry is in crisis. In the year 2012 alone, branded drugs valued at over $30 billion lost patent protection, thereby allowing generic manufacturers to make and sell lower-priced versions of blockbuster drugs [1]. The industry as a whole has been unsuccessful in replacing drugs going off-patent with sufficient new molecular entities (NMEs). Despite staggering investment in R&D – the top ten pharmas spent $60 billion on R&D in 2010 – the number of approvals has changed little over the past decade (Fig. 1.1).

This level of investment without commensurate improvement in approvals of new medicines is likely unsustainable and has, in fact, contributed to waves of mergers in the pharmaceutical industry accompanied by tens of thousands of layoffs in 2007–2012.

Many reasons have been attributed to the lack of apparent productivity in big pharma, but, as the graph in Fig. 1.2 indicates, it is likely that the main culprit is the low probability of success (PoS), perhaps as low as 10%, of investigational drugs entering Phase II clinical trials that are eventually approved [3]. Coupled with the high cost of clinical trials, this low PoS makes drug development a highly risky proposition, and drives the industry to invest in already validated targets that are more likely to yield approvable, albeit less innovative, drugs, at the end of a multi-year effort.

Y. Yao, B. Jallal, K. Ranade (Eds): Genomic Biomarkers for Pharmaceutical Development.
DOI: http://dx.doi.org/10.1016/B978-0-12-397336-8.00001-X

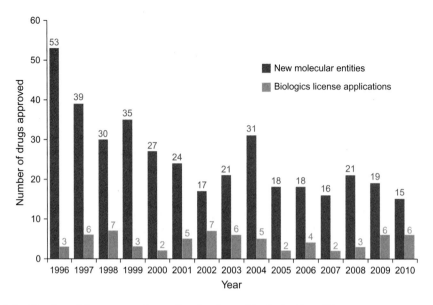

FIGURE 1.1   Number of drug approvals, small molecules and biologics. *From [2].*

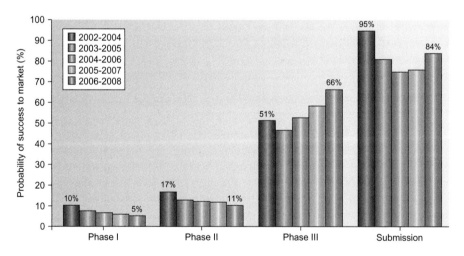

FIGURE 1.2   Probability of success to market from key clinical development milestones. *From [3].*

Paradoxically, it seems that there has never been a better time in biomedical research with all the innovations we are currently witnessing. The Human Genome Project and ancillary efforts, the development of next-generation sequencing and other large scale genome analysis tools have transformed our understanding of diseases, especially cancer [4]. Not a day goes by without news of identification of a 'gene for' a common disease or new insight into a pathway that drives a common cancer. One of the key challenges for pharma R&D will be to translate this explosion in genomic knowledge and new insights into disease pathways into innovative therapeutics that extend and enhance the lives of patients with unmet medical needs.

We believe that judicious application of genomic analysis to develop greater understanding of the molecular underpinnings of complex diseases such as cancer, rheumatic and respiratory diseases will lead to novel targets and therapeutics that are tailored to subsets of these diseases. Together with biomarkers that identify subsets of patients likely to benefit (or not) from targeted therapies, such therapeutics are likely to have a greater PoS in clinical development than those that target all-comers. In this chapter we describe current and emerging translational strategies to apply our expanding genomic knowledge to this end. For the purpose of this discussion, we define 'translational science' as treating the 'right' patient with the 'right' drug' at the 'right dose'. Our objective here is to illustrate broad strategies for identifying the right patient, the right drug and right dose using examples from the literature. We include in our definition of the right patient not only those that will benefit from a novel therapeutic but also those that are less likely to be harmed by it, and we end the chapter with a discussion about adverse drug reactions.

1.2 TWO APPROACHES TO IDENTIFY PATIENT SUBSETS THAT ARE LIKELY TO RESPOND TO INDIVIDUAL THERAPEUTIC INTERVENTIONS

The current paradigm for drug development is to test a new drug candidate in a variety of diseases, such as different types of solid tumors including prostate, breast, colon or hematologic malignancies. Depending on whether a positive signal in a small trial is observed, a couple of indications are selected to conduct follow up larger registrational trials. While this approach has been successful in the past, it is also a key contributor to the ever-increasing cost of drug development; perhaps more importantly it exposes many patients to therapies from which they are unlikely to benefit because the disease of interest is not primarily driven by the targeted pathway in all patients. Viewed from a slightly different perspective, this approach in effect uses the investigational drug to probe the underlying disease in a given patient to assess whether it is amenable to the pathway that is targeted by the drug.

An emerging approach, which is outlined in Fig. 1.3, is to understand, using genomic approaches (e.g., sequencing of tumors or gene expression analysis of relevant tissue), heterogeneity of disease first and thus identify subsets of patients in whom a biological pathway is activated by mutation or by elevated expression of genes in the pathway. The disease in such patients may, therefore, be causatively linked to a particular biological pathway and thus be amenable to therapeutics targeted to the pathway.

Such patients – the right patients – are then targeted with a therapeutic that is tailored to them – the right drug. Key to the success of this approach is a reliable way to identify such patient subsets, i.e., a companion diagnostic to the tailored therapeutic that is economically viable and can be easily implemented in the clinic. We illustrate this approach using the example of crizotinib (Xalkori®) from Pfizer, an ALK (anaplastic lymphoma kinase) and ROS1 (c-ros oncogene1, receptor tyrosine kinase) inhibitor that was approved in 2011 for patients with a particular kind of non-small cell lung cancer (NSCLC) [6].

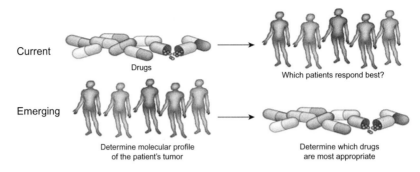

Current

Drugs

Which patients respond best?

Emerging

Determine molecular profile
of the patient's tumor

Determine which drugs
are most appropriate

FIGURE 1.3   Two approaches to drug development: In the past, new molecular entities were tested in a variety of indications, e.g., cancers of different types, to identify those patients most likely to respond. In the emerging translational approach, molecular heterogeneity of a disease is analyzed first, and then therapeutics are developed and tailored to subsets of disease. *Adapted from [5].*

1.2.1 Prospective Analysis: The Case of Crizotinib in NSCLC

The crizotinib story started several years ago, when analysis of a cDNA library from a Japanese patient with lung adenocarcinoma identified a novel fusion between the EML4 and ALK genes with the ability to transform 3T3 fibroblasts [7]. Analysis of a series of biopsies from NSCLC patients revealed that ~5% of patients carry this fusion protein.

Soon after the publication of the initial discovery in 2007, it was found that crizotinib, a small molecule inhibitor of the protein encoded by the ALK gene, was very effective in NSCLC patients whose tumors harbored the ALK fusion gene. It caused tumors to shrink or stabilize in 90% of 82 patients carrying the ALK fusion gene, and tumors shrank at least 30% in 57% of people treated [8]. These promising clinical results led to a Phase II and a Phase III trial, which selectively enrolled NSCLC patients with ALK fusion genes. Astonishingly, within four years of the initial publication by Soda et al., the Food and Drug Administration (FDA) approved crizotinib for the treatment of certain late stage (locally advanced or metastatic) NSCLC patients whose tumors have ALK fusion genes as identified by a companion diagnostic that was approved simultaneously with the drug [6].

There are several important lessons to be learned from the development of crizotinib. First, understanding molecular heterogeneity to identify a mutation or pathway that is causally linked to the disease is crucial to the eventual success. With this knowledge in hand, investigators could design small but highly effective trials targeted to those patients more likely to benefit from the therapy. Such approaches allow drug companies to save both money and time in drug development. The approval for crizotinib was based on two registrational trials that enrolled fewer than 150 subjects each. To better illustrate how targeting patients can improve the PoS of a clinical trial, we performed simulations to estimate the sample sizes that would be required if patients had not been selected in trials of a drug like crizotinib that is targeted to, for instance, only 10% of the population.

Under the assumption of placebo response rates ranging from 6–14% in typical cancer clinical trials [9], if patient randomization is conducted requiring the presence of a biomarker, or biomarker positive group, the minimum sample size needed at 80% power and alpha = 0.05 could be as low as N = 33/arm, with a 30% effect size and 6% response rate in

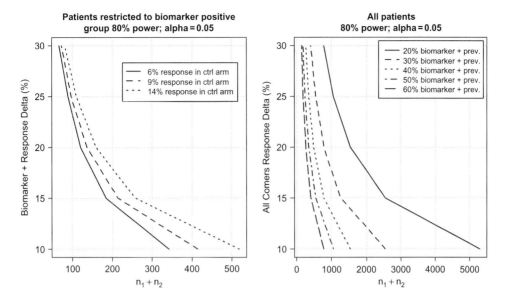

FIGURE 1.4 (left) Relationship between effect size and total sample size when restricting patient inclusion to biomarker positive patients under different control arm response rates (80% power and alpha = 0.05), and (right) the same association showing different levels of biomarker positive patient prevalence, assuming 6% control arm response rate with no restriction to biomarker positive patients, and only the biomarker positive patients showing improvement (80% power and alpha = 0.05). Note that total response delta is plotted on the y axis (right), though sample sizes are calculated using the reduced response delta as explained in the text.

the control arm to as high as N = 259/arm with a 10% effect size and 14% response rate in the control arm [10] (Fig. 1.4 left).

In contrast to this trial design where only biomarker positive patients are included, the sample size requirement in an all-comers trial design, i.e., without selectively enrolling patients, is driven not only by effect size, but by the prevalence of patients with the particular disease sub-type (e.g., NSCLC patients with ALK fusion). For example, assuming a 6% response rate in the control arm, if the trial is not restricted to biomarker positive patients (i.e., those with ALK fusion in this example), and we assume the same effect size in the previous example of 30%, if 10% of the patients are identified as biomarker positive (and only these patients show improvement), the overall improvement rate would be reduced to 3%. Under this design, 1274 patients/arm would be required at alpha = 0.05 and 80% power. If the biomarker positive patient prevalence is identified to be 30%, the reduced effect size is 9% and expected sample size is reduced to 202/arm, under the same assumptions (Fig. 1.4 right). This example illustrates how easily sample size requirements can be affected by either reduced biomarker positive patient prevalence or decreased effect sizes in a clinical trial.

The second lesson from the crizotinib example is the importance of developing strong testable hypotheses early. Although developing robust and reliable hypotheses is often easier said than done, with the right approach equipped with the powerful technologies we currently have, such hypotheses are not out of reach. Fortuitously, in the case of crizotinib one of the clinical sites in enrolling patients in the Phase I trial was already developing tools to assess ALK fusion genes and was able to quickly translate published results into clinical development.

The foregoing discussion has focused on cancer, but similar prospective approaches have been applied to inflammatory diseases as well. As Arron and Harris describe in their chapter on asthma (Chapter 4), gene expression analysis of lung epithelial tissue from treatment-naïve asthma patients revealed that a subset of patients had significantly elevated expression of genes that were regulated by the cytokine IL13. After substantial follow up, a clinical trial demonstrated that this subset of patients, which could be identified with a serum biomarker, derived significant benefit from a novel anti-IL13 therapeutic. This initial observation needs to be confirmed in ongoing Phase III trials, but demonstrates the power of this translational approach. An analogous translational approach to identify a subset of patients with systemic lupus erythematosus is described in the chapter on autoimmune diseases (Chapter 3).

1.2.1.1 Retrospective Analysis to Identify Responders

In contrast to the prospective approaches described above to identify patients who may derive benefit from a therapeutic, we describe below successful examples to identify predictive markers by comparing responders and non-responders, i.e., from retrospective analysis of clinical trials.

1.2.1.2 Large Molecule Inhibitors of Epidermal Growth Factor Receptor (EGFR)

The monoclonal antibodies cetuximab and panitumumab which are targeted against the EGFR have been approved for the treatment of metastatic colorectal cancer. Initial analysis of a small number of responders and non-responders for mutations in genes in the EGFR signaling pathway – KRAS, BRAF, PI3KCA – revealed that KRAS mutations were readily detected in non-responders but not in responders [11]. Of the 11 patients who responded to cetuximab, none was mutant for KRAS; in contrast 13 of 19 non-responders were KRAS mutants. These significant results were confirmed in subsequent large trials of cetuximab [12] and panitumumab [13]. Although the FDA guidance calls for prospective stratification of clinical trials to provide an adequate test of a predictive marker (see Chapter 7), in this case, KRAS mutation status as a predictor of response was approved as a companion diagnostic for cetuximab and panitumumab because of overwhelming evidence from multiple retrospective analyses. Further details of KRAS and response to EGFR targeted therapies can be found in Chapter 2.

1.2.1.3 Small Molecule Inhibitors of EGFR: Case of Gefitinib

The small molecule inhibitors of EGFR, gefitinib (IRESSA™, AstraZeneca) and erlotinib (Tarceva®, Roche) were initially approved in all-comers based on standard registrational trials [14,15]. It was several years post-approval that it was discovered that these inhibitors provide significant benefit to NSCLC patients carrying a particular tumor biomarker, a mutation within the EGFR gene [16–18]. Encouraging anti-tumor activity was observed in NSCLC in a Phase I trial [19]. In two subsequent Phase II trials (IDEAL 1 and IDEAL 2), promising and well-tolerated drug activity was observed. In the same trials, EGFR protein expression levels within the tumor were tested as a potential predictive biomarker for clinical response, but no relationship was found between EGFR protein expression and response [20,21].

In 2004, two independent investigators published retrospective analyses demonstrating that patients with encouraging responses to gefitinib harbored activating mutations in the *EGFR* gene [16,17]. AstraZeneca had already initiated two Phase III trials, ISEL and INTEREST, for gefitinib in unselected NSCLC in 2003 and 2004, respectively. The ISEL study showed some improvement in survival for NSCLC patients treated with gefitinib compared to placebo, but this difference did not reach statistical significance in the overall population. A planned subgroup analysis showed that patients who were female, Asian, non-smokers, or who had adenocarcinomas had better responses [22,23], and demonstrated a trend for increased EGFR gene copy number to be predictive of survival benefit versus placebo [24]. In the INTEREST study, gefitinib demonstrated non-inferiority relative to docetaxel in terms of overall survival with a more favorable tolerability profile and better quality of life. However, there was no evidence from the co-primary analysis to support the hypothesis that patients with high EGFR gene copy number had superior overall survival on gefitinib compared with docetaxel. EGFR mutation-positive patients had longer progression-free survival and higher objective response rate and patients with high EGFR copy number had higher response rates with gefitinib versus docetaxel, but no biomarkers were predictive of differential survival [21,22,25–27].

As part of the testing of patients who had been shown to have a better response to gefitinib versus standard of care chemotherapy, AstraZeneca then conducted a clinical study in an Asian population of non-smokers or light ex-smokers with adenocarcinoma (the IPASS study). IPASS was a randomized, large-scale, double-blind study which compared gefitinib vs. carboplatin/paclitaxel as a first line treatment in advanced NSCLC. IPASS studied 1217 patients, and samples were analyzed for several biomarkers related to the mechanism of action of the drug including EGFR copy number, EGFR expression and EGFR-activating mutations. A pre-planned subgroup analyses showed that progression-free survival (PFS) was significantly longer for gefitinib than chemotherapy in patients with EGFR mutation-positive tumors, and significantly longer for chemotherapy than gefitinib in patients with *EGFR* mutation negative tumors [28; Fig. 1.5]. Although the IPASS study was not designed specifically to test this biomarker hypothesis, this was the first study that was powered to do so; in total, 437 patients in IPASS provided evaluable samples and approximately 60% of these patients carried the EGFR-activating mutations [29].

Thus, in 2009, a targeted monotherapy was able to demonstrate significantly longer PFS than doublet chemotherapy, and the European Commission granted marketing authorization for the treatment with gefitinib of adults with locally advanced or metastatic NSCLC with activating mutations of EGFR-TK, across all lines of therapy [30]. The role of EGFR mutations as a predictive biomarker for response to EGFR TKI therapies has also been demonstrated in other Phase III, randomized clinical trials [31–34].

This validation of biomarkers predictive of response to gefitinib in NSCLC highlights the clinical benefit of identifying patients who are most likely to respond. It also emphasizes the importance of generating robust biomarker hypotheses early in the drug development process. In the absence of strong preclinical and clinical evidence, scientists may explore multiple hypotheses in late stage clinical trials, without the power to test any one biomarker in a rigorous manner.

The development of advanced genome analysis tools, including high throughput next-generation sequencing (NGS), which allows the sequencing of many individual molecules, as opposed to the most prevalent ones in a heterogeneous mixture, has resulted in an

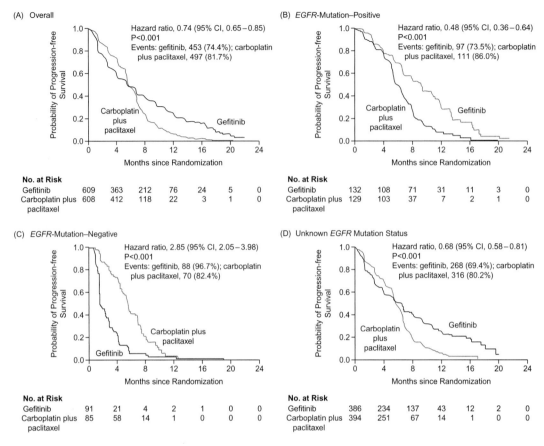

FIGURE 1.5 Kaplan-Meier curves for progression-free survival are shown for the overall population (A), patients who were positive for the *EGFR* mutation (B), patients who were negative for the *EGFR* mutation (C), and patients with unknown *EGFR* mutation status (D). Analyses were performed on the basis of the intention-to-treat population. With respect to the overall population, results of the supportive secondary analyses (including a log-rank test, which is valid under the null hypothesis even when hazards are not proportional, and analysis in the per-protocol population) were consistent with the result of the primary analysis. Hazard ratios were calculated with the use of a Cox proportional-hazards model, with the WHO performance status (0 or 1, or 2), smoking history (nonsmoker or former light smoker), and sex as covariates. EGFR denotes epidermal growth factor receptor.

explosion of surveys of many different tumor types to identify driver mutations and subsets of tumors, and these studies are expected to result in new targeted therapies in the coming years. While the targeted therapies described above demonstrate remarkable response rates in the right patients, eventually patients relapse because their tumors develop resistance. In addition to identifying potentially driver mutations in tumors, NGS has begun to illuminate the cause of such tumor resistance. In a *tour de force* genomic analysis of different regions of a renal cell carcinoma within a single patient, Swanton and colleagues demonstrated significant intra-tumor heterogeneity with regard to mutations [35]. Thus, therapeutics like crizotinib or gefitinib that are targeted to a single mutation could select clones with resistant mutations that are lurking within a tumor at low frequency at the initiation of treatment but become the predominant clone over time. Translating such genomic discoveries about heterogeneity within tumors into tailored therapeutics will be the challenge of the future.

1.2.1.4 *Case of Pegylated Interferon (IFN) in Infectious Disease*

Pegylated IFN in combination with ribavirin is approved for the treatment of chronic infection by the hepatitis C virus. Ge et al. performed a whole genome association study in which they compared the frequencies of almost 600,000 single nucleotide polymorphisms (SNPs) in patients infected with genotype 1 hepatitis C virus between those who had a sustained virologic response to therapy and those that failed to respond [36]. A C/T SNP in the vicinity of the interleukin 28B gene, which encodes the type III interferon IFN-l3, was highly significantly associated with response in whites, blacks and Hispanics. Compared with ~20–40% of patients with a T/T genotype who had a sustained virologic response, 60–80% of those with the C/C genotype had sustained virologic response. In heterozygotes sustained virologic response was intermediate (Hispanics) or similar to that observed in T/T homozygotes (whites and blacks). Subsequent studies also found that this SNP was significantly associated with sustained virologic response [37–39]. Because there were no placebo-treated patients in the study, it could not be determined whether the association between IL28b and sustained virologic response was truly a treatment X genotype interaction effect (as in the KRAS example described above) or whether genotype was a predictor of virologic response even in the absence of treatment. Natural history studies addressed this issue [39,40]. These authors evaluated patients who spontaneously cleared their hepatitis C virus infection, and found that the SNPs associated with response to treatment with interferon-ribavirin were also associated with spontaneous clearance of hepatitis C infection in the absence of treatment, indicating that SNPs in IL28b contribute to variation in virologic response regardless of therapy.

It should be noted that in both cases the analyses that led to the discovery of predictive markers were performed outside of the drug development setting. Because of the large number of tests performed in such analyses (e.g., the genome-wide association tests evaluated hundreds of thousands of SNPs) there is increased likelihood of false-positive associations and initial results need to be replicated in multiple independent studies, as was described for both the KRAS and IL28b examples. Such independent datasets are typically unavailable during the clinical development of a novel therapeutic, making it difficult to interpret such retrospective analyses. In the absence of replication, one runs the risk of investing significant resources in following up potentially spurious associations.

1.3 USE OF A GENOMIC PHARMACODYNAMICS (PD) MARKER TO AID IN DOSE SELECTION IN CLINICAL DEVELOPMENT

The major objectives for first time in human (FTIH) studies for large molecules are usually to evaluate the safety, tolerability, pharmacokinetics and immunogenicity profiles. Due to the long half-life of monoclonal antibodies (mAbs), a single-dose, cohort dose escalation study can take a year or longer to complete. To expedite clinical development and improve the PoS, a relevant, sensitive, and robust PD biomarker that can be easily monitored in trials is of great value for finding the right dose. The PD biomarker can be used to account for differences in target expression and pathway activation in different diseases, and facilitate bridging between clinical trials in different indications. In this section, we present a case where a 'universal' PD marker to measure anti-type I IFN PD effect in multiple autoimmune diseases was developed select doses in two different but related disease indications.

Type I IFNs are a family of cytokines including 14 IFN-α subtypes, IFN-β, -ω, and -κ, and there is strong evidence in the literature that suggests a potential role for type I IFNs in the disease pathogenesis of multiple autoimmune diseases, such as systemic sclerosis (SSc), systemic lupus erythematous (SLE), primary Sjögren's syndrome, and myositis [41–49]. Elevated levels of IFN-α have been detected in the serum of some SLE patients [50,51] and numerous microarray studies have strengthened the hypothesis of type I IFN involvement in the disease pathogenesis of SLE [42,44,52]. Type I IFN-inducible genes can be conveniently measured by techniques such as TaqMan quantitative polymerase chain reaction or microarray with better sensitivity and specificity than traditional protein bioassays [53–56]. Several well defined type I IFN gene signatures have been used to correlate the type I IFN activity with SLE or SSc disease pathogenesis [49,52,57,58] and disease activity [44,59,60], as well as assessing the PD effect of an anti-IFN-α therapy in SLE and myositis [52,61,62].

One large study that enrolled patients who suffered from SLE, dermatomyositis (DM), polymyositis (PM), SSc, or rheumatoid arthritis (RA) showed that a consensus type I IFN signature could be identified in the whole blood (WB) of these patients [49; Fig. 1.6]. Among these, a five-gene type I IFN-inducible gene panel was developed to identify subpopulations with concordant activation of the type I IFN pathway between the peripheral blood and disease-affected tissues in SLE, DM, PM and SSc [49; Fig. 1.7]. This development allowed the potential utility for the same type I IFN gene panel as a PD marker for anti-type I IFN receptor (IFNAR) therapy in both SLE and SSc.

Although multiple clinical trials evaluating anti-IFN-α monoclonal antibody in SLE have been conducted and larger confirming trials are underway, it remains to be seen whether targeting IFNAR, which suppresses all type I IFN subtypes, might show a favorable benefit risk ratio in SLE. It also remains to be seen whether neutralizing the type I IFN signature, a surrogate of suppressing the type I IFN pathway, shows positive correlation with clinical benefit in SLE. MEDI-546, a fully human IgG1κ mAb, is one such drug that targets the subunit 1 of the IFNAR1. It was evaluated in a FTIH study in subjects with SSc (MI-CP180) where a complete neutralization of the type I IFN signature was observed (Fig. 1.8). The pharmacokinetic (PK) profile and type I IFN signature were analyzed using a mechanistic model incorporating the binding of MEDI-546 to IFNAR1, internalization kinetics of MEDI-546/IFNAR1 complex as determined from confocal imaging studies, and the inhibition of type I IFN signature by MEDI-546 [63].

To support the transition of MEDI-546 from a FTIH study in SSc patients to a large proof of concept (PoC) study in SLE, we used translational simulations to bridge across the two patient populations in lieu of an additional Phase I trial in SLE. Briefly, stochastic simulations suggested that a 300 mg monthly fixed dose could sustain suppression of the type I IFN signature in a typical SLE subject; a higher dose (1000 mg) was also recommended for the PoC trial to ensure adequate drug exposure and target neutralization in skin, especially for SLE patients with substantially elevated type I IFN signature at baseline (Fig. 1.9). In addition, the 1000 mg dose would ensure against potential divergence of the SLE simulation assumptions (e.g., the potency and activity of MEDI-546 in SLE could be different from that

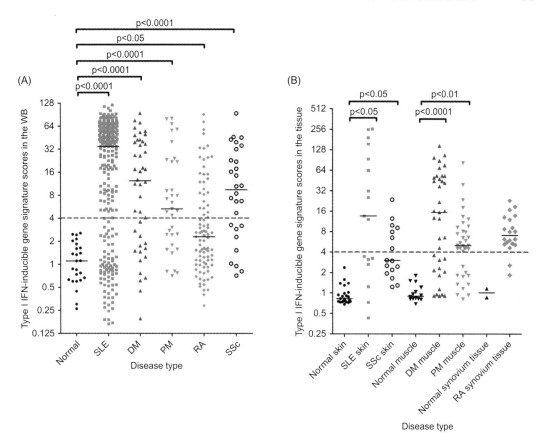

FIGURE 1.6 Overexpression of type I IFN-inducible genes as captured by the five gene type I IFN signature scores in the WB and disease tissues of subjects with SLE, DM, PM, RA and SSc. Five gene type I IFN signature scores in (A) WB and (B) disease tissues for healthy and diseased subjects [healthy subjects (N = 24); SLE (N = 262); DM (N = 44); PM (N = 33); RA (N = 89); SSc (N = 38)]. The p values for the signature scores between healthy subjects and SLE, DM, PM, RA and SSc subjects in the WB after adjustment for gender and age are $p < 0.0001$, $p < 0.0001$, $p < 0.0001$, $p < 0.05$ and $p < 0.0001$, respectively. The p values for the lesional skin from SLE (N = 16) and SSc (N = 16) subjects compared with normal healthy subjects (N = 25) is $p < 0.05$ for both and $p < 0.0001$ and $p < 0.01$ in muscle specimens between DM (N = 37) or PM (N = 36) subjects and healthy subjects (N = 14), respectively. No statistical test was calculated in the RA comparison due to sample size (N = 2 vs. N = 20), although the trend is apparent. Horizontal bars represent the median values for each group and the gray dashed line indicates the threshold for signature positive or negative status.

in SSc). The population modeling using a relevant PD marker and stochastic simulations greatly facilitated the bridging across different treatment periods, dosing methods, and patient populations of the FTIH and the PoC studies.

The clinical development of a new drug is a lengthy and costly process with low odds of success. Contrary to common impression, the clinical development of biotherapeutics is not quicker or cheaper than that of small molecule drugs [64,65]. The early clinical development of biotherapeutics, in particular Phase I, is much lengthier than for small molecules,

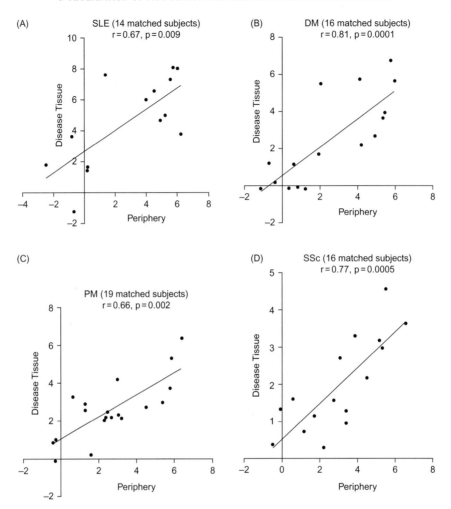

FIGURE 1.7   Correlation scatter plots of the five gene type I interferon signature scores between that in disease tissue (y axis) and the whole blood (x axis) in subjects with (A) systemic lupus erythematosus (SLE, N = 14), (B) dermatomyositis (DM, N = 16), (C) polymyositis (PM, N = 19) and (D) systemic scleroderma (SSc, N = 16). Spearman rank correlation coefficients and p values are provided to quantify the association for each disease plot.

which typically require single-center Phase I study in healthy volunteers. In the absence of a definitive efficacy signal from early phase studies in patients, a sensitive, disease-relevant, and robust biomarker can greatly aid the interpretation of clinical results. This case provided an interesting study to develop novel, robust genomic PD biomarkers for a therapeutic that targets the type I IFN signaling, which in conjunction with population PK-PD modeling and stochastic simulations allowed smooth transition of MEDI-546 from SSc to SLE. This approach not only provided crucial results to avoid additional dose-finding Phase I or Phase IIa studies and associated extra development costs, but also provided patients, an early potential opportunity to participate in a Phase IIb study.

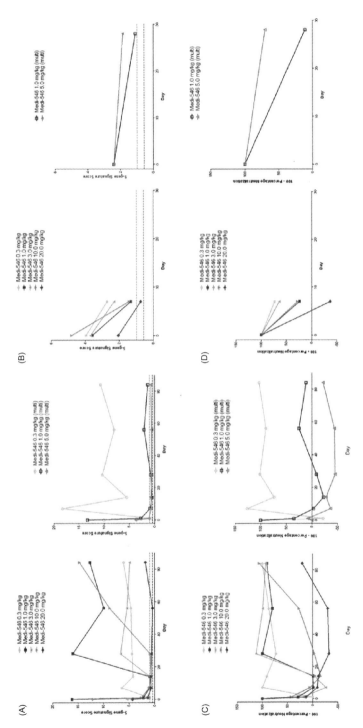

FIGURE 1.8 Median type I IFN (A and B) signature profiles as calculated by the five gene panel and (C and D) percent remaining in SSc patients following single or multiple IV administrations of MEDI-546 in (A and C) whole blood specimens or (B and D) skin specimens from MI-CP180 trial. For each pair of plots, the single and multiple dose treatment cohorts have been separated into their respective graph. X axis represents time from the start of the study in days, where day 0 is pre-treatment. Target modulation for each dose cohort is reported as a percentage from starting values of 100%, so each point post treatment for each cohort indicates the median percentage of remaining type I IFN signature. Only type I IFN score positive SSc patients at baseline were plotted. The minimum and average type I IFN signature scores among the pool of normal healthy controls are shown as black and red dashed lines respectively (A and B). Based on the calculation of percent remaining type I IFN signature, when the signature reaches values below the signature of the average of the normal control, the target modulation can result in a negative value, as is seen in certain cohorts in C and D. *Courtesy of Clin. Pharmacol. Ther.*

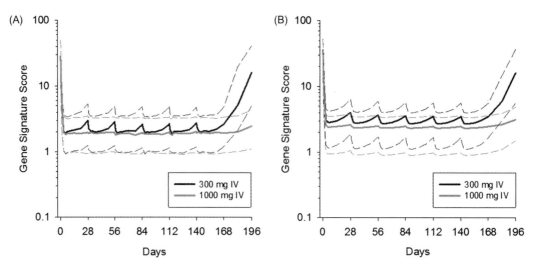

FIGURE 1.9   Simulated type I IFN signature profiles in peripheral blood (A) and skin tissue (B) of SLE patients upon multiple IV administrations of MEDI-546 (fixed dose) once every four weeks. The solid lines represent the medians of 1000 simulated profiles while dotted lines represent the lower or upper quartiles. The observed upper boundaries (mean + 2 standard deviations) of the type I IFN gene signature in the blood and skin of healthy donors were 2.9 and 1.8, respectively. *Courtesy of Clin. Pharmacol. Ther.*

1.4 SAFETY

Adverse drug reactions (ADRs) which are defined by the World Health Organization (WHO) as: 'harmful, unintended reactions to medicines that occur at doses normally used for treatment' [66] are among the top ten reasons for hospitalization in the United States [67]. Indeed ADRs have resulted in the withdrawal of a number of drugs and multi-billion dollar lawsuits in recent years. The causes of ADRs and how to prevent them have been the topic of numerous commentaries [68]. Some of the preventable causes of ADRs include medication errors including wrong medication, inappropriate dose or drug-drug interaction. While the economic cost of ADRs and how to reduce them is a large subject, in this chapter we focus on how translational approaches can help to understand the underlying mechanism and how this understanding in turn can lead to strategies to reduce the likelihood of an ADR.

A common reaction to the question, 'How safe should a drug be?' is typically 'Extremely safe. There should be no severe ADR.' While this reaction is understandable, no drug is without risk and in the context of drug development it is more useful to assess the value of any therapeutic to society in terms of its benefit:risk ratio. This ratio is not a constant; rather it can vary by disease area, and even over time. For example, statins which are used to reduce cholesterol can be considered to have a high benefit:risk ratio, as they have demonstrated impressive benefit in reducing cardiovascular morbidity and mortality over two decades and are associated with very rare ADRs such as rhabdomyolysis with a frequency of one event in 10,000 [69]. In contrast, cancer therapies typically have low response rates and significant ADRs associated with them, so any new therapeutic that confers incremental benefit over standard of care even when associated with significant ADRs can have an acceptable benefit:risk ratio. For inflammatory diseases like RA, the benefit:risk ratio can considered to be intermediate between these two extremes. The reason is that monoclonal antibodies against tumor necrosis factor have proven to be very efficacious and have become the standard of care. Despite their obvious success these therapeutics provide clinically significant benefit in only up to 50–60% of patients leaving a significant proportion of the RA population underserved [70]. Moreover, anti-TNFs are associated with serious ADRs including malignancy and tuberculosis. Another interesting example is provided by anti-retrovirals that are used to control the human immunodeficiency virus (HIV). Up to the mid-1990s when HIV and the ensuing AIDS were often fatal, significant ADRs caused by anti-retrovirals were considered acceptable because those drugs were the only proven means of controlling HIV. With the introduction of anti-retroviral cocktails, HIV has become a chronic and manageable condition in Western countries which in turn has significantly improved the benefit:risk ratio, thus resulting in less acceptance of ADRs for new anti-HIV medications.

Previous sections of this chapter described proactive translational strategies to identify patients more likely to benefit from a therapeutic. It is challenging, however, to employ such approaches to ADRs because they are treatment-emergent and usually unpredictable. We describe below two genomic approaches that have been used to understand the mechanism of an ADR. This mechanistic knowledge has been applied to develop a test to screen for patients susceptible to ADR caused by the anti-retroviral abacavir or to develop follow-on compounds to muraglitazar, an investigational anti-diabetic.

1.4.1 Case Study: Abacavir in HIV

Abacavir (ABC) is a nucleotide analogue that competitively inhibits HIV reverse transcriptase. ABC can be associated with an ADR characterized by fever, malaise, gastrointestinal symptoms and internal organ involvement in approximately 5–8% of patients who begin therapy with the drug [71]. The syndrome can be accompanied by a mild-to-moderate rash in 70% of patients with ABC hypersensitivity and is associated with severe hypotension and possible death upon re-challenge, in contrast to the complete resolution of symptoms 72 hours after withdrawal of the drug [72].

Two groups independently performed a case-control genetic association study between HLA alleles and susceptibility to ABC-induced hypersensitivity and found highly significant association with the HLA Class I allele, HLA-B*5701 [73,74]. These studies demonstrate better than any simulation that when the genetic effect size is substantial (odds ratios for the associated allele were 24 and 103) even relatively small case-control studies – DNA from 18 and 85 cases of ABC-induced hypersensitivity was evaluated in the two studies – can provide useful information. These initial reports were confirmed by several other groups and led to a prospective trial to test whether screening for HLA-B*5701 could reduce the incidence of ABC-induced hypersensitivity. PREDICT-1 was a large (N = 1956) double-blind study that randomized patients to either receive real-time HLA-B*5701 screening and exclusion of ABC for those positive for HLA-B*5701 or ABC initiation and clinical monitoring with retrospective HLA-B*5701 analysis [75]. The study demonstrated that HLA-B*5701 had 100% sensitivity and 97% specificity for ABC HSR and a negative predictive value of 100% and a positive predictive value of 47%; i.e., not only are patients without the HLA-B*5701 allele highly unlikely to develop ABC-induced hypersensitivity, but about half those who carry the HLA-B*5701 are also unlikely to develop this ADR. A case-control study extended these findings from a study that enrolled predominantly white subjects in the US, and demonstrated 100% sensitivity of the HLA-B*5701 allele for ABC-induced hypersensitivity in white and black subjects [76]. These studies provided strong evidence for the clinical utility of HLA-B*5701 to prevent ABC HSR, generalizable across race. Another important aspect of the PREDICT-1 study was that 45% of HLA-B*5701 carriers were able to tolerate ABC. The study of ABC-exposed HLA-B*5701-positive subjects holds the key to understanding the additional factors required for the development of the syndrome in hypersensitive patients or, conversely, the protective factors in tolerant patients. These efforts, together with the availability of a test to screen for HLA-B*5701 culminated in the following FDA warning in the package insert for abacavir:

> 'Patients who carry the HLA-B*5701 allele are at high risk for experiencing a hypersensitivity reaction to abacavir. Prior to initiating therapy with abacavir, screening for the HLA-B*5701 allele is recommended; this approach has been found to decrease the risk of hypersensitivity reaction. Screening is also recommended prior to reinitiation of abacavir in patients of unknown HLA-B*5701 status who have previously tolerated abacavir. HLA-B*5701-negative patients may develop a suspected hypersensitivity reaction to abacavir; however, this occurs significantly less frequently than in HLA-B*5701-positive patients.' [77].

This example underscores the many hurdles, not least of which is prospective testing of a candidate patient selection/exclusion tool, that need to be overcome before an exciting genomic result can be translated into clinical practice to aid patients.

1.4.2 PPARgamma Agonists in Type II Diabetes

We turn now to a translational approach that employed identifying risk factors as above but then applied the knowledge pre-clinically to a drug discovery effort to identify new pipeline molecules with reduced liability of causing the ADR associated with the lead compound. Peroxisome-proliferator activated receptor (PPAR) agonists that activate the transcription factor PPARgamma have been approved to control glycemia in Type II diabetics [78]. These agents are frequently associated with edema or fluid retention and less commonly congestive heart failure [78]. Muraglitazar was an investigational drug that activated both PPARalpha and PPARgamma transcription factors. In Phase II clinical trials it showed robust efficacy with regard to lowering blood glucose levels and was associated with dose-dependent edema [79]. By comparing subjects who developed edema to those that did not, Geese and colleagues identified renin, endothelin-1 and the beta-adrenergic receptor as potential susceptibility genes for muraglitazar-induced edema [80]. In follow up gene expression analysis in a cell-culture model, it was found that renin and endothelin-1 were significantly induced by muraglitazar and all other known PPAR agonists. Remarkably, there was a correlation between the potency of a PPAR agonist to induce expression of these genes in the cell-culture model and its propensity to cause edema in humans. The authors applied this knowledge to three other PPAR alpha/gamma agonists that were indistinguishable in their ability to reduce blood glucose levels in an animal model to predict one candidate drug that would be less likely to cause edema. Consistent with this prediction, it was found that among the three candidate drugs the least potent molecule with regard to inducing expression of renin and endothelin was also the one least likely to cause edema in a non-human primate model. This example is a powerful illustration of how careful analysis of even early clinical trials can translate to knowledge that can be applied to improve the discovery pipeline.

1.5 CONCLUSIONS

Genomics research has transformed our understanding of disease and led to tangible benefits in drug discovery and development. In particular, the landscape of cancer has changed based on our evolving understanding of driver mutations and heterogeneity of disease. In turn, this knowledge has led to accelerated drug development timelines as shown (Fig. 1.10).

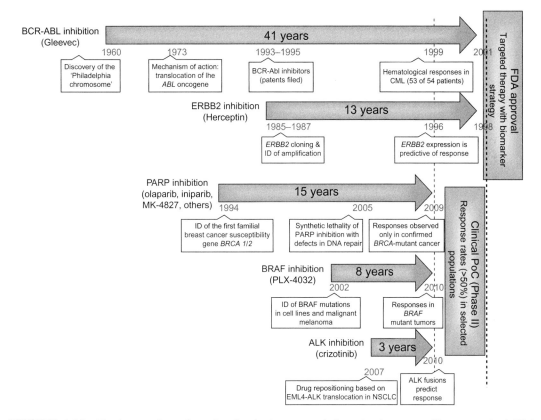

FIGURE 1.10 The historical timelines for developing targeted therapies in cancer. Gleevec received FDA approval 41 years after the discovery of the Philadelphia chromosome mutation and hyperactive BCR-ABL protein in chronic myelogenous leukemia (CML). By contrast, the more recent discovery of chromosomal rearrangements (translocations) of ALK in NSCLC has rapidly translated into registration trials and approval for crizotinib. Likewise, the development paradigm for selective BRAF inhibitors, as exemplified by PLX4032, underlines the much faster pace of translation (8 years, compared with Gleevec or Herceptin) once the driver status (in this case BRAF mutations) had been established for an indication (malignant melanoma). The FDA approval of Herceptin and the accompanying diagnostic test for HER2 expression (HercepTest) proved the value of biomarker-driven trials that are informed by mechanistic insights gained from cancer genetics. The functional understanding of DNA-repair mechanisms, and the role of BRCA1 and BRCA2 mutations in sensitizing tumors to PARP inhibition, inform current registration trials of PARP inhibitors in BRCA-associated cancer types and patients that carry the BRCA mutation. *Adapted from Chin et al., Courtesy of Nature Medicine.*

In summary, we have described broad strategies to identify the right drug at the right dose for the right patient. By capitalizing on technological improvements in genomics, we believe that the future of drug development is bright so long as we put the patient first, i.e., understand the molecular heterogeneity within a disease and then develop therapeutics that are targeted to particular subsets of patients. This translational approach, in turn, is likely to increase the probability of success of clinical trials, a pre-requisite for a sustainable pharma business.

References

[1] Thomas K. Generic drug makers see a drought ahead. New York Times 2012;December 4:B1.
[2] Allison M. Reinventing clinical trials. Nat Bio 2012;30:41–9.
[3] Arrowsmith J. A decade of change. Nat Rev Drug Discov 2012;11:17–18.
[4] Chin L, Andersen JN, Futreal PA. Cancer genomics: from discovery science to personalized medicine. Nat Med 2011;17:297–303.
[5] Yap TA, Sandhu SK, Workman P, de Bono JS. Envisioning the future of early anti-cancer drug development. Nat Rev Cancer 2010;7:514–23.
[6] Xalkori Package Insert. <Xalkori.com>, accessed February 2013.
[7] Soda M, Choi YL, Enomoto M, Takada S, Yamashita Y, Ishikawa S, et al. Identification of the transforming EML4-ALK fusion gene in non-small cell lung cancer. Nature 2007;448:561–6.
[8] Ou S-HI. Crizotinib: a novel and first-in-class multitargeted tyrosine kinase inhibitor for the treatment of anaplastic lymphoma kinase rearranged non-small cell lung cancer and beyond. Drug Des, Dev Ther 2011;5:471–85.
[9] Chvetzoff G, Tannock IF. Placebo effects in oncology. J Natl Cancer Inst 2003;95:19–29.
[10] Simon R. The use of genomics in clinical trial design. Clin Cancer Res 2008;14:5984–93.
[11] Lièvre A, Bachet JB, Le Corre D, Boige V, Landi B, Emile JF, et al. KRAS mutation status is predictive of response to cetuximab therapy in colorectal cancer. Cancer Res 2006;66:3992–5.
[12] Karapetis CS, Khambata-Ford S, Jonker DJ, O'Callaghan CJ, Tu D, Tebbutt NC, et al. K-ras mutations and benefit from cetuximab in advanced colorectal cancer. N Engl J Med 2008;359:1757–65.
[13] Amado RG, Wolf M, Peeters M, Van Cutsem E, Siena S, Freeman DJ, et al. Wild-type KRAS is required for panitumumab efficacy in patients with metastatic colorectal cancer. J Clin Oncol 2008;26:1626–34.
[14] Cohen MH, Williams GA, Sridhara R, Chen G, McGuinn Jr WD, Morse D, et al. United States Food and Drug Administration Drug Approval summary: Gefitinib (ZD1839; Iressa) tablets. Clin Cancer Res 2004;10:1212–18.
[15] Johnson JR, Cohen M, Sridhara R, Chen YF, Williams GM, Duan J, et al. Approval summary for erlotinib for treatment of patients with locally advanced or metastatic non-small cell lung cancer after failure of at least one prior chemotherapy regimen. Clin Cancer Res 2005;11:6414–21.
[16] Lynch TJ, Bell DW, Sordella R, Gurubhagavatula S, Okimoto RA, Brannigan BW, et al. Activating mutations in the epidermal growth factor receptor underlying responsiveness of non-small-cell lung cancer to gefitinib. N Engl J Med 2004;350:2129–39.
[17] Paez JG, Jänne PA, Lee JC, Tracy S, Greulich H, Gabriel S, et al. EGFR mutations in lung cancer: correlation with clinical response to gefitinib therapy. Science 2004;304:1497–500.
[18] Pao W, Miller V, Zakowski M, Doherty J, Politi K, Sarkaria I, et al. EGF receptor gene mutations are common in lung cancers from 'never smokers' and are associated with sensitivity of tumors to gefitinib and erlotinib. Proc Natl Acad Sci U S A 2004;101:13306–11.
[19] Kris MG, Herbst R, Rischin D, LoRusso P, Baselga J, Hammond L, et al. Objective regressions in non-small cell lung cancer patients treated in Phase I trials of ZD1839 ('Iressa'), a selective tyrosine kinase inhibitor that blocks the epidermal growth factor receptor (EGFR). Lung Cancer 2000;2972(Suppl. 1) poster 233.
[20] Kris MG, Natale RB, Herbst RS, et al. Efficacy of gefitinib, an inhibitor of the epidermal growth factor receptor tyrosine kinase, in symptomatic patients with non-small cell lung cancer. a randomized trial. JAMA 2003;290:2149–58.
[21] Fukuoka M, Yano S, Giaccone G, et al. Multi-institutional randomized Phase II trial of gefitinib for previously treated patients with advanced non-small-cell lung cancer. J Clin Oncol 2003;21:2237–46.

[22] Thatcher N, Chang A, Parikh P, Rodrigues Pereira J, Ciuleanu T, von Pawel J, et al. Gefitinib plus best supportive care in previously treated patients with refractory advanced non-small-cell lung cancer: results from a randomised, placebo controlled, multicenter study (Iressa Survival Evaluation in Lung Cancer). Lancet 2005;366:1527–37.

[23] Chang A, Parikh P, Thongprasert S, et al. Gefitinib (IRESSA) in patients of Asian origin with refractory advanced non-small cell lung cancer: subset analysis from the ISEL study. J Thorac Oncol 2006;1:847–55.

[24] Hirsch FR, Varella-Garcia M, Bunn Jr PA, Franklin WA, Dziadziuszko R, Thatcher N, et al. Molecular predictors of outcome with gefitinib in a Phase III placebo-controlled study in advanced non-small-cell lung cancer. J Clin Oncol 2006;24:5034–42.

[25] Kris MG, Natale RB, Herbst RS, Lynch Jr TJ, Prager D, Belani CP, et al. Efficacy of gefitinib, an inhibitor of the epidermal growth factor receptor tyrosine kinase, in symptomatic patients with non-small cell lung cancer: a randomized trial. JAMA 2003;290:2149–58.

[26] Kim ES, Hirsh V, Mok T, Socinski MA, Gervais R, Wu YL, et al. Gefitinib versus docetaxel in previously treated non-small-cell lung cancer (INTEREST): a randomised Phase III trial. Lancet 2008;372:1809–18.

[27] Douillard JY, Shepherd FA, Hirsh V, Mok T, Socinski MA, Gervais R, et al. Molecular predictors of outcome with gefitinib and docetaxel in previously treated non-small-cell lung cancer: data from the randomized Phase III INTEREST trial. J Clin Oncol 2010;28:744–52.

[28] Mok TS, Wu YL, Thongprasert S, Yang CH, Chu DT, Saijo N, et al. Gefitinib or carboplatin-paclitaxel in pulmonary adenocarcinoma. N Engl J Med 2009;361:947–57.

[29] Fukuoka M, Wu YL, Thongprasert S, Sunpaweravong P, Leong SS, Sriuranpong V, et al. Biomarker analyses and final overall survival results from a Phase III, randomized, open-label, first-line study of gefitinib versus carboplatin/paclitaxel in clinically selected patients with advanced non-small-cell lung cancer in Asia (IPASS). J Clin Oncol 29; 2866–74.

[30] http://www.iressa.com/_mshost5259502/content/6906430/July2009.pdf

[31] Maemondo M, Inoue A, Kobayashi K, Sugawara S, Oizumi S, Isobe H, North-East Japan Study Group. Gefitinib or chemotherapy for non-small-cell lung cancer with mutated EGFR. N Engl J Med 2010;362:2380–8.

[32] Mitsudomi T, Morita S, Yatabe Y, Negoro S, Okamoto I, Tsurutani J, West Japan Oncology Group. Gefitinib versus cisplatin plus docetaxel in patients with non-small-cell lung cancer harboring mutations of the epidermal growth factor receptor (WJTOG3405): an open label, randomised Phase III trial. Lancet Oncol 2010;11:121–8.

[33] Rosell R, Carcereny E, Gervais R, Vergnenegre A, Massuti B, Felip E, Spanish Lung Cancer Group in collaboration with Groupe Français de Pneumo-Cancérologie and Associazione Italiana Oncologia Toracica. Erlotinib versus standard chemotherapy as first-line treatment for European patients with advanced EGFR mutation-positive non-small-cell lung cancer (EURTAC): a multicenter, open-label, randomised Phase III trial. Lancet Oncol 2012;13:239–46.

[34] Zhou C, Wu YL, Chen G, Feng J, Liu XQ, Wang C, et al. Erlotinib versus chemotherapy as first-line treatment for patients with advanced EGFR mutation-positive non-small-cell lung cancer (OPTIMAL, CTONG-0802): a multicenter, open-label, randomised, Phase III study. Lancet Oncol 2011;12:735–42.

[35] Gerlinger M, Rowan AJ, Horswell S, Larkin J, Endesfelder D, Gronroos E, et al. Intratumor heterogeneity and branched evolution revealed by multiregion sequencing. N Engl J Med 2012;366:883–92.

[36] Ge D, Fellay J, Thompson AJ, Simon JS, Shianna KV, Urban TJ, et al. Genetic variation in IL28B predicts hepatitis C treatment-induced viral clearance. Nature 2009;461:399–401.

[37] Tanaka Y, Nishida N, Sugiyama M, Kurosaki M, Matsuura K, Sakamoto N, et al. Genome-wide association of IL28B with response to pegylated interferon-alpha and ribavirin therapy for chronic hepatitis C. Nat Genet 2009;41:1105–9.

[38] Suppiah V, Moldovan M, Ahlenstiel G, Berg T, Weltman M, Abate ML, et al. IL28B is associated with response to chronic hepatitis C interferon-alpha and ribavirin therapy. Nat Genet 2009;41:1100–4.

[39] Rauch A, Kutalik Z, Descombes P, Cai T, Di Iulio J, Mueller T, Swiss Hepatitis C Cohort Study; Swiss HIV Cohort Study. Genetic variation in IL28B is associated with chronic hepatitis C and treatment failure: a genome-wide association study. Gastroenterology 2010;138:1338–45.

[40] Thomas DL, Thio CL, Martin MP, Qi Y, Ge D, O'Huigin C, et al. Genetic variation in IL28B and spontaneous clearance of hepatitis C virus. Nature 2009;461:798–801.

[41] Theofilopoulos AN, Baccala R, Beutler B, Kono DH. Type I interferons (alpha/beta) in immunity and autoimmunity. Annu Rev Immunol 2005;23:307–36.

[42] Bennett L, Palucka AK, Arce E, Cantrell V, Borvak J, Banchereau J, et al. Interferon and granulopoiesis signatures in systemic lupus erythematosus blood. J Exp Med 2003;197:711–23.

[43] Niewold TB, Rivera TL, Buyon JP, Crow. MK. Serum type I interferon activity is dependent on maternal diagnosis in anti-SSA/Ro-positive mothers of children with neonatal lupus. Arthritis Rheum 2008;58:541–6.

[44] Baechler EC, Batliwalla FM, Karypis G, Gaffney PM, Ortmann WA, Espe KJ, et al. Interferon-inducible gene expression signature in peripheral blood cells of patients with severe lupus. Proc Natl Acad Sci USA 2003;100:2610–15.

[45] Greenberg SA, Pinkus JL, Pinkus GS, Burleson T, Sanoudou D, Tawil R, et al. Interferon-alpha/beta-mediated innate immune mechanisms in dermatomyositis. Ann Neurol 2005;57:664–78.

[46] Farina G, Lafyatis D, Lemaire R, Lafyatis. R. A four-gene biomarker predicts skin disease in patients with diffuse cutaneous systemic sclerosis. Arthritis Rheum 2010;62:580–8.

[47] Assassi S, Mayes MD, Arnett FC, Gourh P, Agarwal SK, McNearney TA, et al. Systemic sclerosis and lupus: points in an interferon-mediated continuum. Arthritis Rheum 2010;62:589–98.

[48] Hjelmervik TO, Petersen K, Jonassen I, Jonsson R, Bolstad. AI. Gene expression profiling of minor salivary glands clearly distinguishes primary Sjogren's syndrome patients from healthy control subjects. Arthritis Rheum 2005;52:1534–44.

[49] Higgs BW, Liu Z, White B, Zhu W, White WI, Morehouse C, et al. Patients with systemic lupus erythematosus, myositis, rheumatoid arthritis and scleroderma share activation of a common type I interferon pathway. Ann Rheum Dis 2011;70:2029–36.

[50] Hua J, Kirou K, Lee C, Crow MK. Functional assay of type I interferon in systemic lupus erythematosus plasma and association with anti-RNA binding protein autoantibodies. Arthritis Rheum 2006;54:1906–16.

[51] Pascual V, Farkas L, Banchereau J. Systemic lupus erythematosus: all roads lead to type I interferons. Curr Opin Immunol 2006;18:676–82.

[52] Yao Y, Richman L, Higgs BW, Morehouse CA, de los Reyes M, Brohawn P, et al. Neutralization of interferon-alpha/beta-inducible genes and downstream effect in a Phase I trial of an anti-interferon-alpha monoclonal antibody in systemic lupus erythematosus. Arthritis Rheum 2009;60:1785–96.

[53] Tannous BA, Kim DE, Fernandez JL, Weissleder R, Breakefield XO. Codon-optimized Gaussia luciferase cDNA for mammalian gene expression in culture and *in vivo*. Mol Ther 2005;11:435–43.

[54] Dall'era MC, Cardarelli PM, Preston BT, Witte A, Davis Jr. JC. Type I interferon correlates with serological and clinical manifestations of SLE. Ann Rheum Dis 2005;64:1692–7.

[55] Kirou KA, Lee C, George S, Louca K, Papagiannis IG, Peterson MG, et al. Coordinate overexpression of interferon-alpha-induced genes in systemic lupus erythematosus. Arthritis Rheum 2004;50:3958–67.

[56] Yao Y, Higgs BW, Morehouse C, de Los Reyes M, Trigona W, Brohawn P, et al. Development of potential pharmacodynamic and diagnostic markers for anti-IFN-alpha monoclonal antibody trials in systemic lupus erythematosus. Hum Genomics Proteomics 2009, http://dx.doi.org/pii:374312.

[57] Coelho LF, de Oliveira JG, Kroon EG. Interferons and scleroderma—a new clue to understanding the pathogenesis of scleroderma? Immunol Lett 2008;118:110–15.

[58] Eloranta ML, Franck-Larsson K, Lovgren T, Kalamajski S, Ronnblom A, Rubin K, et al. Type I interferon system activation and association with disease manifestations in systemic sclerosis. Ann Rheum Dis 2010;69:1396–402.

[59] Bilgic H, Ytterberg SR, Amin S, McNallan KT, Wilson JC, Koeuth T, et al. Interleukin-6 and type I interferon-regulated genes and chemokines mark disease activity in dermatomyositis. Arthritis Rheum 2009;60:3436–46.

[60] Baechler EC, Bauer JW, Slattery CA, Ortmann WA, Espe KJ, Novitzke J, et al. An interferon signature in the peripheral blood of dermatomyositis patients is associated with disease activity. Mol Med 2007;13:59–68.

[61] Higgs BW, Zhu W, Morehouse C, White WI, Brohawn P, Guo X, et al. Sifalimumab, an anti-IFN-α mAb, shows target suppression of a type I IFN signature in blood and muscle of dermatomyositis and polymyositis patients. Ann Rheum Dis 2013;February 23, http://dx.doi.org/10.1136/annrheumdis-2012-202794.

[62] Merrill JT, Wallace DJ, Petri M, Kirou KA, Yao Y, White WI, et al. Safety profile and clinical activity of sifalimumab, a fully human anti-interferon alpha monoclonal antibody, in systemic lupus erythematosus: a Phase I, multicenter, double-blind randomised study. Ann Rheum Dis 2011;70:1905–13.

[63] Wang B, Higgs BW, Chang L, Vainshtein I, Liu Z, Streicher K, et al. Pharmacogenomics and translational simulations to bridge indications for an anti-interferon-α receptor antibody. Clin Pharmacol Ther 2013;February 14. http://dx.doi.org/:10.1038/clpt.2013.35.J.

[64] DiMasi A, Grabowski. HG. Economics of new oncology drug development. J Clin Oncol 2007;25:209–16.

[65] DiMasi JA, Feldman L, Seckler A, Wilson. A. Trends in risks associated with new drug development: success rates for investigational drugs. Clin Pharmacol Ther 2010;87:272–7.

[66] World Health Organization. <http://www.who.int/mediacentre/factsheets/fs293/en/index.html.> Medicines: safety of medicines – adverse drug reactions. 2008, accessed January 2013.

[67] Lazarou J, Pomeranz BH, Corey PN. Incidence of adverse drug reactions in hospitalized patients: a meta-analysis of prospective studies. JAMA 1998;15:1200–5.

[68] Agency for Healthcare Research and Quality. <http://www.ahrq.gov/qual/aderia/aderia.htm> Reducing and Preventing Adverse Drug Events to Reduce Hospital Costs. Accessed January 2013.

[69] SEARCH Collaborative Group Link E, Parish S, Armitage J, Bowman L, Heath S, et al. SLCO1B1 variants and statin-induced myopathy – a genomewide study. N Engl J Med 2008;359:789–99.

[70] Abbvie. <http://www.rxabbvie.com/pdf/humira.pdf>. Humira package insert. Accessed February 2013.

[71] Cutrell AG, Hernandez JE, Fleming JW, Edwards MT, Moore MA, Brothers CH, et al. Updated clinical risk factor analysis of suspected hypersensitivity reactions to abacavir. Ann Pharmacother 2004;38:2171–2.

[72] Shapiro M, Ward KM, Stern JJ. A near-fatal hypersensitivity reaction to abacavir: case report and literature review. AIDS Read 2001;11:222–6.

[73] Hetherington S, Hughes AR, Mosteller M, Shortino D, Baker KL, Spreen W, et al. Genetic variations in HLA-B region and hypersensitivity reactions to abacavir. Lancet 2002;359:1121–2.

[74] Mallal S, Nolan D, Witt C, Masel G, Martin AM, Moore C, et al. Association between presence of HLA-B*5701, HLA-DR7, and HLA-DQ3 and hypersensitivity to HIV-1 reverse-transcriptase inhibitor abacavir. Lancet 2002;359:727–32.

[75] Mallal S, Phillips E, Carosi G, Molina JM, Workman C, Tomazic J, PREDICT-1 Study Team. HLA-B*5701 screening for hypersensitivity to abacavir. N Engl J Med 2008;358:568–79.

[76] Saag M, Balu R, Phillips E, Brachman P, Martorell C, Burman W, Study of Hypersensitivity to Abacavir and Pharmacogenetic Evaluation Study Team. High sensitivity of human leukocyte antigen-b*5701 as a marker for immunologically confirmed abacavir hypersensitivity in white and black patients. Clin Infect Dis 2008;46:1111–18.

[77] GlaxoSmithKline. <http://www.viivhealthcare.com/~/media/Files/G/GlaxoSmithKline-Plc/pdfs/us_ziagen.pdf>. Ziagen/abacavir package insert. Accessed January 2013.

[78] Stein SA, Lamos EM, Davis SN. A review of the efficacy and safety of oral antidiabetic drugs. Expert Opin Drug Saf 2013;12:153–75.

[79] Rubin CJ, Viraswami-Appanna K, Fiedorek FT. Efficacy and safety of muraglitazar: a double-blind, 24-week, dose-ranging study in patients with type 2 diabetes. Diab Vasc Dis Res 2008;6:205–15.

[80] Geese WJ, Achanzar W, Rubin C, Hariharan N, Cheng P, Tomlinson L, et al. Genetic and gene expression studies implicate renin and endothelin-1 in edema caused by peroxisome proliferator-activated receptor gamma agonists. Pharmacogenet Genomics 2008;18:903–10.

2

Personalized Health Care (PHC) in Cancer

Nicholas C. Dracopoli[1], Katie Streicher[2]

[1]Janssen Research and Development, LLC Radnor, Pennsylvania
[2]MedImmune, LLC Gaithersburg, Maryland

2.1 INTRODUCTION

Empirical drug development strategies in oncology are becoming unsustainable. Still, only ~10% of first in class oncology drugs entering first in human (FIH) testing eventually obtain regulatory approval [1]. The reasons for this high attrition rate are complex but often result from poor *in vitro* and *in vivo* models which lead to early failure in clinical development, or because a poor understanding of the underlying molecular pathology of the disease leads to the development of the drug in suboptimal indications. Novel approaches to personalized healthcare are required to reduce the attrition of novel oncology drugs in development and to improve outcome for cancer patients.

The discovery and application of novel biomarkers is essential for the successful implementation of personalized healthcare for cancer patients. These biomarkers are used for many purposes, including validating the mechanism of action (MOA) of a new drug, exploring pharmacokinetic/pharmacodynamic (PK/PD) interactions, predicting response to therapy, identifying mechanisms of resistance to therapy, and determining prognosis and disease outcome. However, converting biomarkers to clinical practice has been difficult. There are 21,403 biomarkers listed in the GVK BIO Online Biomarker (GOBIOM) database (January 31, 2013), but only 32 of these are listed as valid biomarkers by the Food and Drug Administration (FDA). All the clinically validated markers are single analyte biomarkers, and no complex molecular profiles that predict response to therapy have yet been approved by the FDA.

The following sections describe successful examples of personalized healthcare in oncology and can be seen as precedents for future development of first in class oncology drugs. These represent only the first steps using simple biomarkers that measure the status of the drug target or pathway to predict patient response. Today, a total of 12 drugs have been approved in oncology with predictive biomarkers. However, these approved drugs only

Y. Yao, B. Jallal, K. Ranade (Eds): Genomic Biomarkers for Pharmaceutical Development.
DOI: http://dx.doi.org/10.1016/B978-0-12-397336-8.00002-1

23

target six different genes in several tyrosine kinase signaling pathways. Successful implementation of personalized healthcare in oncology will require different strategies to develop complex multi-analyte biomarkers for drugs with more diverse mechanisms of action.

2.2 BIOMARKERS

The term biomarker has been used in multiple contexts, but one definition put forth by the FDA is: 'a measurable characteristic that is an indicator of normal biologic processes, pathogenic processes, and/or response to a therapeutic or other intervention'. Multiple types of biomarkers that meet this definition are currently used, with many more being developed, to identify groups of patients who have a better prognosis, are more likely to respond to a particular treatment, are successfully responding to their treatment, or may go on to develop resistance. As such, pharmacodynamic, predictive, prognostic, resistance, and surrogate biomarkers will be discussed, along with current examples relevant to the oncology field.

2.2.1 Pharmacodynamic Biomarkers

Pharmacodynamic biomarkers measure the effects of a drug on the tumor itself and/or the disease in general. They can be used to guide dose selection in the early stages of clinical development. A general example is the use of the extent of proliferation or apoptosis in a tumor following administration of a particular drug, or the change in a downstream factor that is regulated by the drug target, such as changes in protein phosphorylation after inhibition of a protein kinase. A specific example of using a pharmacodynamic biomarker in a clinical setting relates to the use of imatinib in chronic myelogenous leukemia (CML). Imatinib blocks the oncogenic fusion protein break point cluster-Abelson tyrosine kinase (BCR-ABL) from initiating a signaling cascade important for the development of CML [2]. The reduction in protein kinase activity of BCR-ABL is used to indicate response to imatinib in patients with CML, such that the magnitude of inhibition correlates with clinical outcome. This pharmacodynamic relationship has been used successfully to benefit patients with CML by identifying the most effective clinical doses, which are considerably lower than the doses associated with imatinib toxicity [2,3].

2.2.2 Predictive Biomarkers

Predictive biomarkers assess the likelihood that tumors will respond to a particular treatment. They can be used to guide the choice of therapy for a particular group of patients and may also play a role in identifying optimal drug dose and/or predicting toxicity. Multiple examples of the practical use of predictive biomarkers exist in clinical oncology. There are many types of predictive markers, which can be grouped into the following categories: target/pathway expression; target/pathway mutation or genetic alteration; and drug metabolism. The type of predictive marker identified not only speaks to its MOA, but also relates to the timing of when the particular marker is most likely to be incorporated into clinical development. For example, if the expression of the drug target itself predicts response to treatment, there is great potential that this hypothesis was studied early in the drug development process and could be validated in prospective trials and co-developed as a companion diagnostic. Similarly for mutations/genetic alterations, if they exist in the drug target itself, the opportunity for co-development of a companion diagnostic may accelerate drug approval; however, if the genetic alteration occurs outside of the target pathway, this may only be identified through retrospective genetic analyses of responder/non-responder populations following completed large Phase II and Phase III trials. Finally, discovery of alterations in drug metabolism due to a patient's genotypic background may be extremely difficult to predict prior to large Phase III clinical trials or even prior to FDA approval due to the need for retrospective analyses of large patient populations to identify/validate such relationships, unless pre-clinical work has implicated a drug-metabolizing enzyme that is inhibited or is involved in activating or inactivating a small molecule drug.

2.2.2.1 *Target/Pathway Expression*
ESTROGEN RECEPTOR (ER) AND HORMONE THERAPY IN BREAST CANCER

One of the original predictive biomarkers shown to have clinical relevance is the estrogen receptor (ER) in breast cancer (BC). In the late 1800s, Sir George Beatson reported that three women with advanced BC experienced regression of their tumors following oophorectomy, starting the idea that hormonal manipulation could influence BC progression [4]. Subsequent trials demonstrated that sensitivity to hormonal manipulation was observed in tumors that express ER alpha [5]. Once the connection between ER expression and response to hormonal therapy had been established, it became critical to develop a method to identify ER expressing tumors. Originally this was done using ligand binding assays; however, current practice is to use immunohistochemistry (IHC) in formalin-fixed, paraffin embedded (FFPE) tumor blocks to determine ER status [6]. Several systems have been devised to score results from IHC tests of ER in breast tissue, one of which is the Allred score [7]. The Allred score provides an overall score of either 0 or 2–8 by summing the proportion of cells with positive staining and the average staining intensity [7]. To determine a cutoff distinguishing ER-positive samples from ER-negative samples, the IHC assay results were compared with clinical outcome. This comparison revealed that an Allred score of 3 was the lowest score predicting a positive response to hormonal therapy (Fig. 2.1) and this score corresponds to as few as 1–10% positive tumor cells [7,8]. The American Society of Clinical Oncology (ASCO) and the College of American Pathologists incorporated this information into specific guidelines for testing and scoring of ER expression, such that ≥1% of tumor cells staining positive is considered sufficient to predict response to hormonal therapy [8]. Using the criteria of 1–10% ER expression as positive, approximately 70% of all breast cancers qualify for hormonal therapy [8].

Support for the hypothesis that ER levels correlate with benefit from anti-estrogen therapy emerged in the early 1970s, in studies showing that 50% of ER+ patients with advanced BC experienced regression following endocrine ablative therapy or hormone therapy, whereas ER− patients rarely responded to these therapies [9]. In 1998, a meta-analysis of 55 clinical trials comparing tamoxifen, a selective ER modulator that blocks transcription of estrogen-dependent genes and inhibits breast tumor proliferation, versus placebo for the adjuvant treatment of early invasive BC showed a 50% reduction in disease recurrence rates and a 28% reduction in mortality rate in ER+ patients following 5 years of tamoxifen treatment, whereas little benefit was observed in patients with ER− cancers [10]. Another recent update of this meta-analysis

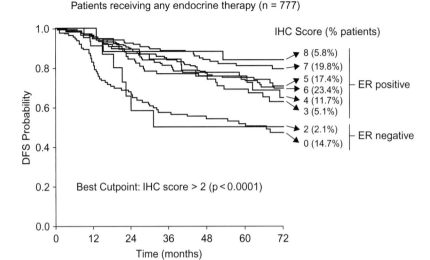

FIGURE 2.1 Comparison of disease-free survival curves for all possible Allred IHC scores for ER staining used to determine cutoff score distinguishing a positive from a negative result [8]. ER positivity is indicated by an Allred score ≥3 (indicating >1% of cells with weak staining).

utilized 194 randomized trials where patients received adjuvant chemotherapy and tamoxifen, and showed that tamoxifen reduced the 15-year breast cancer recurrence rate from 45% to 33% and reduced BC mortality by 35% in ER+ patients, but was ineffective in ER− patients [11]. Based on these data and others, the use of tamoxifen as hormonal therapy is currently the standard of care in breast cancer, especially in premenopausal women. Other therapies that inhibit the ER exist, such as aromatase inhibitors, which prevent the conversion of androgens to estrogens, causing a reduction in estradiol levels [6]. Regardless of the estrogen-modulating therapy, clear responses are only seen in patients who meet the criteria for ER positivity.

Despite the benefits observed following tamoxifen and other hormonal therapies, additional work needs to be done to identify patients likely to respond to treatments that inhibit ER function. The negative predictive value (NPV) of ER expression is high, meaning that ER-negative patients almost never benefit from hormone therapy; however, the positive-predictive value (PPV) of ER expression is lower, in that only about 50% of ER+ patients respond to hormone therapy [11–13]. Additionally, approximately 33% of cancers treated with adjuvant hormonal therapy will recur [6], and there are currently no tests to prospectively identify which ER+ patients will ultimately respond or develop resistance to endocrine therapy. The development of additional biomarkers to be used in combination with ER status to identify patients more likely to respond to anti-estrogen therapy hopefully will be the next step in the evolution of the treatment of ER+ cancer.

HERCEPTIN/TRASTUZUMAB FOR BREAST CANCER

HER2 (also known as ErbB2) is a member of the epidermal growth factor receptor family, which regulates key cellular pathways, including proliferation, migration, and adhesion [14]. HER2 is overexpressed in approximately 25% of breast tumors, leading to poorer prognosis for these patients [14–17]. Amplification of the gene for HER2, detected by fluorescence *in situ* hybridization (FISH) occurs in approximately the same proportion [18]. Trastuzumab/ Herceptin is a recombinant humanized mAb that targets HER2 protein and prevents downstream signaling of this pathway [18]. Treatment with trastuzumab as monotherapy leads to responses in 15–35% of patients with the response rate increasing to 50–70% when it is administered with chemotherapy to patients with increased expression or amplification of HER2 in their tumors, termed HER2+ [19–23]. The pivotal trial for this drug in combination with chemotherapy enrolled 469 patients with previously untreated, HER2+, metastatic breast cancer [18]. A central laboratory reviewed the tumor samples to quantify the level of HER2 overexpression using IHC staining, which is scored semi-quantitatively as 2+ for weak to moderate staining of the entire tumor cell membrane and 3+ for more than moderate immunostaining [18]. Trastuzumab was associated with an increase in the objective response rate from 32% to 50%, longer median survival, and a 20% reduction in the risk of death. Retrospective analysis of HER2 expression by IHC and copy number differences by FISH revealed that the antibody was most active in patients with 3+ HER2 staining intensity or HER2 gene amplification. No patient with 2+ HER2 IHC staining had a response unless gene amplification was detected by FISH.

Due to the increased response rate in HER2+ patients, multiple clinical tests have been approved for use in tumor biopsies to measure HER2 protein levels or gene amplification or other changes in HER2 activation in order to select patients more likely to respond to this therapy [15–18,24]. All HER2 tests have been validated by retrospective analyses and were not used to qualify patients for enrollment in studies leading to the initial approval of trastuzumab by the FDA. HER2 testing has continued to evolve, with many clinical laboratories using both FISH and IHC to evaluate HER2 [18]. IHC remains the most frequent initial test for HER2 status and is performed on approximately 80% of newly diagnosed breast cancers in the US. Advantages of IHC testing include its broad availability, relatively low cost, and easy preservation of stained slides, with disadvantages including lack of a positive internal control and difficulties interpreting the semi-quantitative subjective scoring system of 1+, 2+, 3+ [16]. On the other hand, FISH has a more objective scoring system and an internal control due to known levels of the HER2 gene in cells that do not have gene amplification. FISH can be more expensive than IHC and does not allow for long term preservation and storage of slides, but is considered to be more reproducible [25–27]. Currently, testing for HER2 status often starts with screening by IHC, where results of 0 and 1+ are considered negative, 2+ is considered equivocal and referred for FISH testing, and 3+ is considered positive [16]. Although this is the most typical testing algorithm in clinical practice, numerous tests using other methodologies have been approved to measure HER2 status in BC (Table 2.1). Unfortunately, clear understanding of how to interpret the results from each of these tests and utilize them to impact treatment decisions is still lacking [24]. The ability to correctly identify HER2+ patients is important because patients misclassified as HER2− will be denied potential benefit from anti-HER2 therapies and those misclassified as HER2+ may endure potentially toxic and expensive therapy with little chance of clinical benefit.

Although the issues with HER2 testing results are important, the more critical issue may be that the NPV of the HER2 test is high, but the PPV is only around 25–40% [18,28,29], indicating that about 60–70% of HER2 positive patients do not respond to HER2 targeted therapy. With a low PPV of measuring HER2 alone (by any test) and resistance developing

TABLE 2.1   Examples of Tests to Measure HER2 Status in Breast Cancer [24]

Company Location	Name of Test Status	Technology
Biogenex	InSite HER2/neu CB11 FDA approved	Immunohistochemistry assay using a monoclonal antibody directed against the internal domain of HER2/neu
Dako	HER2 FISH pharmDx Kit FDA approved	FISH assay to determine HER2 gene amplification in FFPE breast cancer specimens. Gene amplification = the ratio between the number of signals from the hybridization of the HER2 gene probe and the number of signals from the hybridization of the reference chromosome 17 probe
Dako	HercepTest FDA approved	Semi-quantitative IHC assay to determine HER2 protein overexpression in breast cancer tissues routinely processed for histological evaluation
Genomic Health	Oncotype DX CLIA validated	RT PCR-based assay analyzing expression of a panel of 21 genes, among them HER2. Oncotype DX predicts disease recurrence and assesses benefit from certain types of chemotherapy
Invitrogen	SPOT-Light HER2 CISH Kit FDA approved	Chromogenic in situ hybridization (CISH) using a DNA probe. Quantifiable results are visualized under a standard brightfield microscope
Monogram Biosciences	HERmark Breast Cancer Assay CLIA-validated	Proximity-based assay, which provides direct quantitative measurements of HER2 total protein and HER2 homodimer levels
Siemens Healthcare Diagnostics	HER2/neu ELISA FDA approved	Sandwich enzyme immunoassay using mouse monoclonal for capture and a different biotinylated mouse monoclonal antibody for the detection of human HER2/neu protein. Detection is by direct chemiluminescence. Protein is quantified by spectrophotometry
Ventana-RocheTucson	Inform HER2 Silver in situ Hybridization Approved in Europe but not by FDA	Fully automated silver in situ hybridization assay for HER2 and chromosome 17 detection. Chromogenic signals are detected through the use of silver deposition technology. Results and morphological significance can be interpreted using conventional brightfield microscopy
Ventana-Roche	Pathway anti-HER2/neu (Clone CB11) FDA approved	Semiquantitative immunohistochemistry assay using a monoclonal antibody for the detection of c-erbB-2 (HER2) antigen using Ventana's family of automated instrument platforms
Vysis (Abbott)	PathVysion HER2 DNA Probe Kit FDA approved	Fluorescence in situ hybridization (FISH) assay to determine HER2 amplification, using LSI HER2 probe, which spans HER2, and CEP 17 probe, which hybridizes to the alpha satellite DNA located at the centromere of the chromosome

GENOMIC BIOMARKERS FOR PHARMACEUTICAL DEVELOPMENT

rapidly in nearly all patients [16,18,28], there is certainly room for improvement in predicting patient response to anti-HER2 therapies. Measuring other markers in addition to HER2 may greatly increase the ability to predict response to anti-Her2 therapies like trastuzumab, which would be of considerable benefit to patients with breast cancer.

PD-1 FOR IMMUNOTHERAPY IN MULTIPLE INDICATIONS

The programmed death-1 (PD-1) T cell co-receptor and its ligands B7-H1/PD-L1 and B7-DC/PD-L2 play important roles in maintaining an immunosuppressive tumor microenvironment. The major role of PD-1 is to limit the activity of T cells in the periphery during an inflammatory response [30]. PD-L1 is commonly upregulated on many tumor types, where it inhibits local T cell responses, while PD-1 is expressed on the majority of tumor-infiltrating lymphocytes [30]. Encouraging clinical results have validated this pathway as a target for cancer immunotherapy [31]. Clinical activity has been observed in patients with melanoma, renal cell carcinoma, colorectal cancer (CRC), and non-small-cell lung cancer (NSCLC) [31]. Antitumor activity of the anti-PD-1 antibody BMS-936558 was observed at all doses tested, and in melanoma the response rates ranged from 19 to 41% in each dose cohort. Tumor cell surface expression of the PD-1 ligand PD-L1 in pre-treatment FFPE tumor biopsies emerged as a biomarker of response, consistent with known biology. 25 of the 42 patients examined were positive for PD-L1 expression by IHC; of the 25 patients, 9 had an objective response (36%) compared to 0 of the 17 PD-L1 negative patients that experienced an objective response [31]. Approximately 1 in 5 individuals treated with the anti-PD-1 antibody BMS-936558 had durable responses to the drug [31] and it is clear that the expression of PD-L1 in tumors is a candidate predictive marker that warrants further exploration for use with this promising immunotherapy.

2.2.2.2 Target/Pathway Mutation or Genetic Alteration

EGFR MAB THERAPY IN CRC

The epidermal growth factor receptor (EGFR) is frequently overexpressed and/or mutated in cancers, including CRC, NSCLC, and glioblastoma [32–34]. Prevention of ligand binding to this receptor inhibits signal transduction pathways such as RAS/RAF/MAPK and PI3K/AKT cascades, which promote cell growth proliferation, invasion, and metastasis [35]. Both monoclonal antibodies (mAbs; cetuximab and panitumumab) and small molecule tyrosine kinase inhibitors (TKIs; gefitinib and erlotinib) are approved therapies for these cancers [32–34,36–38]. Trials with EGFR mAbs showed responses for these agents to be between 10 and 30% (single agent mAb and plus chemotherapy, respectively) in chemorefractory patients with metastatic CRC [35]. The low response rate and the poor correlation of EGFR expression with response to anti-EGFR therapies led to the development of alternative biomarkers. KRAS is a gene involved in multiple EGFR-mediated pathways, and mutations in this gene lead to constitutively activated MAPK or PI3K signaling independent of EGFR activation. KRAS mutations occur in a large proportion of metastatic CRCs (30–50%) and are negative prognostic factors, which supported the hypothesis that KRAS mutations could play a role in patient response to anti-EGFR treatment [35]. Retrospective analysis of several randomized clinical trials revealed that patients with mutations in the KRAS gene rarely benefited from treatment with EGFR mAbs, making KRAS status an important

TABLE 2.2   FDA Criteria for Retrospective Biomarker Analyses

The trial must be adequate, well-conducted and well-controlled.

Tumor tissue must be obtained in ≥95% of the registered and randomized study subjects and an evaluable result (presence of wild-type or mutant KRAS) must be available for ≥90% of the registered and randomized study subjects.

Before analysis, the FDA must have reviewed the assay methodology and determined that it has acceptable analytical performance characteristics (for example, sensitivity, specificity, accuracy, precision) under the proposed conditions for clinical use.

Before analysis of clinical outcomes based on the genetic testing, agreement with the FDA must be reached on the analytical plan and promotional claims.

Sample size must be sufficiently large to be likely to ensure random allocation to each of the study arms for factors that were not used as stratification variables for randomization.

Genetic analysis must be performed according to the qualified assay method by individuals who are masked to treatment assignment and clinical outcome results.

predictive biomarker for poor response to EGFR mAbs [35,39–41]. Although post hoc evaluation of clinical data is not generally accepted by the FDA, it acknowledged in this case that incorporating this predictive marker into routine clinical practice would benefit patients by sparing them treatment unlikely to be efficacious. Accordingly, the FDA has updated the labels of EGFR mAbs to indicate that administration is not recommended in patients with mutant KRAS. Additionally, the European Medicines Agency (EMEA) approved panitumumab contingent on the requirement for determining patient KRAS status prior to administration [39,40,42]. This situation of retrospective analyses leading to additional biomarker discovery is not unique to EGFR mAbs, and as such, the FDA has provided guidance regarding the criteria that need to be met before it will consider the results of these types of analyses (Table 2.2). This is encouraging for other therapies that may be able to find similar relationships and reflects the overall understanding that there can be potential value in retrospective analyses, provided the conditions are appropriate.

EGFR TKI THERAPY IN NSCLC

As described for CRC, the EGFR is also expressed in NSCLC where it has been correlated with aggressive disease and decreased survival [35]. Similar to what was seen in CRC with EGFR mAbs, EGFR expression does not correlate with response to EGFR TKIs. In this case, however, mutations in the target itself, EGFR, were shown to predict response to EGFR TKIs in NSCLC [43–46]. The majority of these mutations were observed in two hotspots: in frame deletions in exon 19 and a point mutation involving the replacement of leucine with arginine at codon 858 (L858R) in exon 21 [33,44]. EGFR activating mutations result in ligand-independent EGFR signaling, leading to uncontrolled tumor proliferation. EGFR gene copy number has also been associated with response to TKIs, but with lower sensitivity and specificity than mutations [33]. Together, clinical studies have shown that 55–82% of patients with EGFR activating mutations responded to TKIs, compared with a 20–30% response rate to platinum-based chemotherapy [46–50]. In contrast, those lacking EGFR activating mutations rarely benefit from TKIs [46–50]. Specifically, in a trial by the West Japan Oncology

TABLE 2.3 EGFR TKIs Compared to Chemotherapy in Unselected and Selected Patient Populations

Patient Selection	EGFR TKIs			Platinum Doublets		
	RR (%)	TTP or PFS (months)	OS (months)	RR (%)	TTP or PFS (months)	OS (months)
Unselected	20–23	3–4.4	13.9–14.7	17–21	3.1–4.2	7.4–8.1
Molecularly Selected*	62–71	9.2–13.1	28.0	32–47	5–6	23.6

(Adapted from Lopez-Chavez and Giaccone [43].)
Abbreviations: OS = overall survival; PFS = progression free survival; RR = response rate; TTP = time to progression.
*EGFR TKI sensitizing mutation positive.

group, the EGFR TKI gefitinib was shown to be superior to conventional chemotherapy for the first line treatment of NSCLC with sensitizing EGFR mutations [47]. In this study, gefitinib with cisplatin plus docataxel was given to 172 chemotherapy-naïve patients with stage IIIB/IV NSCLC who had EGFR mutations that were either exon 19 deletions or L858R point mutations. Patients treated with gefitinib had a longer progression-free survival than chemotherapy-treated patients and the objective response rate was 62.1% in the gefitinib group compared to 32.2% in the chemotherapy group [47]. Prior to this trial, EGFR TKIs had been tested in unselected patient populations with a frequency of EGFR mutations at 12–15% and these patients achieved a response rate of 20–23%. This is substantially lower than that observed in patients selected for EGFR mutations, highlighting the value of this predictive marker. Table 2.3 illustrates the difference in response rate and overall survival achievable with EGFR TKIs when targeting the appropriate patient population.

BCR/ABL TRANSLOCATION, ALK FUSIONS, AND PML-RARA FUSION

Chronic myeloid leukemia accounts for 15% of newly diagnosed cases of leukemia in adults [51]. Most CML cases are associated with a translocation between the Abelson murine leukemia (ABL) gene on chromosome 9 with the breakpoint cluster region (BCR) gene on chromosome 22, resulting in the fusion protein BCR-ABL acting as a constitutively active tyrosine kinase found only in cancer cells [2,3]. Small molecule TKIs were developed to target this abnormally expressed BCR-ABL protein in CML cells, leading to a dramatic change in the management of this disease, and improving the 10-year overall survival from approximately 20% to 80–90% [52]. Measurement of the BCR-ABL protein can be accomplished by measuring the co-localization of genomic probes specific to the BCR and ABL genes using FISH or by amplifying the region around the splice junction between BCR and ABL using polymerase chain reaction (PCR) [51]. The existence of BCR-ABL predicts response to imatinib, a selective TKI of this fusion protein that has demonstrated substantial and durable responses in CML, leading to FDA approval in 2001 [2,3,53].

Crizotinib is a small molecule TKI originally synthesized to inhibit activated HGFR (c-Met); however, studies also showed that it could inhibit anaplastic lymphoma kinase (ALK) [54–56]. This is clinically significant due to the role of ALK in a subtype of non-Hodgkin's lymphoma (fusion of the ALK gene with the NPM gene leads to constitutively active ALK) and NSCLC (fusion between ALK and EML4 leads to constitutively active ALK). The EML4-ALK fusion causes ligand-independent dimerization of the ALK kinase domain, leading to constitutive proliferation and inhibition of apoptosis [57]. The EML4-ALK fusion is found in 2–7% of unselected NSCLC patients, but this number can increase to approximately 13% in patients who are female, Asian, never/light smoker, and have an adenocarcinoma histology to their tumor [58]. Although there is a relatively low percentage of NSCLC patients with this genetic alteration, a 57% overall response rate to treatment with crizotinib was observed in a cohort of 82 patients with ALK rearrangements, as detected by FISH, with virtually no responses in the absence of ALK abnormalities [54–56,58–60]. Additionally, in individuals matched for clinical characteristics, ALK-positive patients receiving crizotinib had an improved two-year overall survival (57% compared to 36%) compared to ALK-positive patients who did not receive crizotinib [55]. FDA approval of crizotinib occurred in 2011, with accelerated approval due to the existence of a prospective genomic biomarker and parallel companion diagnostic development [61]. Required for this approval was that ALK testing be performed using a FDA approved test; therefore, the Vysis ALK break-apart FISH probe kit (Abbott Molecular) was approved for the detection of ALK translocation-positive NSCLC in parallel with the approval of crizotinib.

Acute promyelocytic leukemia (APL) accounts for 10% of all acute myeloid leukemias and more than 99% of cases have a fusion between the retinoic acid receptor and the PML gene (PML-RARa fusion) [62–65]. The detection of a PML-RARa t(15,17) translocation is a diagnostic biomarker for APL and this genetic change functions to repress retinoic acid responsive target genes. Its presence correlates with response to ATRA therapy (all trans retinoic acid) followed by arsenic trioxide, which targets the fusion protein for degradation [62,64,65].

BRAF MUTATION FOR VEMURAFENIB

BRAF is a serine-threonine kinase that activates the MAP/ERK kinase signaling pathway [66]. In 2002, an important advance in the understanding of v-raf murine sarcoma viral oncogene homolog B1 (BRAF) function came from the discovery that the BRAF gene is mutated in many different cancers [67]. Most mutations occur at codon 600, replacing valine most typically with glutamic acid (V600E mutation) and leading to significant increases in kinase activity that drive cancer cell proliferation [67]. This mutation occurs in about half of all melanomas [68] and has varying prevalence in other cancers [67]. Vemurafenib is a BRAF inhibitor synthesized in early 2002 possessing mild selectivity for BRAF V600E over the wild type enzyme in biochemical assays, but a pronounced selectivity for mutant BRAF compared to wild type in inhibiting proliferation of melanoma cell lines [67]. Before initiating clinical studies, work was initiated to develop a companion diagnostic to identify melanoma patients with a BRAF mutation. Roche Diagnostics developed a real-time PCR assay to detect the BRAF V600E mutation in FFPE tissue samples that had exceptional analytical performance (125ng genomic DNA containing 5% mutant alleles produced >96% hit rate) providing greater sensitivity and specificity over conventional sequencing [69]. In Phase I studies of vemurafenib, 32 patients with metastatic melanoma treated with 960mg of vemurafenib twice daily had an incredible 81% response rate in patients with V600E mutation [66,67]. In Phase II and Phase III trials, the BRAF V600E diagnostic assay was used as an enrollment criterion [69]. Results

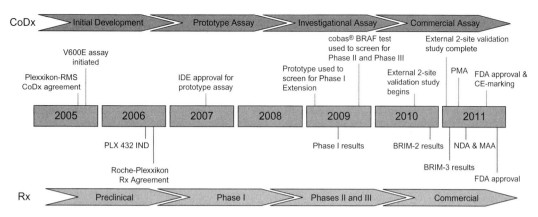

FIGURE 2.2 Milestones in the co-development of vemurafenib and the cobas® 4800 BRAF V600 Mutation Test. *(Reprinted from Lopez-Rios et al. [72].)*

showed a 53% response rate in a 132 patient Phase II study in individuals with melanoma positive for the BRAF V600E mutation [70] and a >50% overall response rate in a 675 patient Phase III study comparing vemurafenib with dacarbazine in individuals with previously untreated or unresectable stage IIIc or stage IV melanoma positive for the BRAF V600E mutation [71]. All clinical data, including the results from the diagnostic assay, were submitted to the FDA in May of 2011, with the concurrent approval of the diagnostic assay and vemurafenib in August 2011. The overall time period of less than five years (see Fig. 2.2 for timeline of co-development of vemurafenib and the BRAF V600E diagnostic) between the filing of the investigational new drug application and the FDA approval of vemurafenib highlights the key value of identifying predictive biomarkers to select the right drug for the right patient.

2.2.2.3 Drug Metabolism

UGT1A AND IRINOTECAN

Irinotecan is a topoisomerase I inhibitor used to treat several solid tumor types, including CRC, in combination with other chemotherapeutic agents. Inhibition of topoisomerase I by irinotecan and its active metabolite, SN-38, prevents re-ligation of single-stranded DNA breaks induced during cellular replication [73,74]. Uridine diphosphate glucuronosyltransferase 1A1 (UGT1A1) is a hepatic enzyme primarily responsible for conjugation of bilirubin and endogenous hormones [73,74]. UGT1A1 also catalyzes the glucuronidation of SN-38, the active metabolite of irinotecan and the main source of treatment-related toxicity [73–75]. A recent meta-analysis demonstrated that genetic variation in UGT1A1 is moderately predictive of severe irinotecan-induced hematologic toxicity and neutropenia at intermediate doses of irinotecan and strongly predictive at high doses (>250 mg/m^2), but at low doses these patients have a comparable incidence of toxicity compared to other patients [73–76]. Therefore, knowledge of a patient's UGT1A1 polymorphism status could guide the selection of appropriate starting dosages, reducing the risk of severe toxicity and improving the chances that therapy can be maintained. The most studied of the UGT1A1 polymorphisms is UGT1A1*28, which results in reduced UGT1A1 activity, affecting the elimination of drug substrates [75]. When the *28 allele is present on one chromosome, a 25% decrease in enzyme activity can be observed, and when the allele is homozygous, the activity is reduced by 70% [77]. In 2005, the FDA recommended that the package insert of irinotecan be amended to warn of the elevated risk of neutropenia for patients with specific UGT1A1 genotypes [76]. Additional studies are warranted to further understand the precise relationship between UGT1A1 genotype and dose of irinotecan, but an example of how this information could be used in clinical decision making for CRC is shown in Fig. 2.3 (reprinted from [75]).

TPMT AND CHEMOTHERAPY IN ALL

About 80% of children with acute lymphoblastic leukemia (ALL) can be cured with combination chemotherapy [78]. Treatment-related toxicity can be life-threatening and is the number one reason for discontinuing chemotherapy. Mercaptopurine interferes with the activity of DNA-processing enzymes due to structural changes in DNA after incorporation of thioguanine nucleotides (TGNs) [78,79]. Mercaptopurine metabolic conversion is in competition with methylation by thiopurine methyltransferase (TPMT), such that variant alleles of TPMT lead to increases in TGN concentration and a higher risk of hematopoietic toxicity after mercaptopurine treatment [78–81]. Lowering doses of mercaptopurine in TPMT heterozygotes and in TPMT-deficient patients allows the administration of the full doses of other chemotherapies while maintaining TGN concentrations that are comparable to patients treated with high doses of mercaptopurine [80,81]. Although patients with non-functional TPMT alleles have reduced tolerance to mercaptopurine, they can be safely treated with lower doses, illustrating a clear rationale for assessing TPMT genotype before initiating therapy. Currently, ALL protocols are being adjusted for TPMT genotype, permitting all patients to receive treatment without high toxicity and without losing efficacy [78–81].

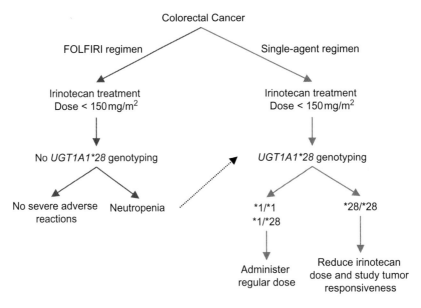

FIGURE 2.3   Algorithm for the use of *UGT1A1* genotyping in clinical decision making for colon cancer.

CYP2D6 AND HORMONE THERAPY IN BREAST CANCER

Breast cancers expressing the ER are dependent on estrogen for growth. Selective ER modulators inhibit estrogen binding to ERs, reducing the proliferation of ER+ tumors. Tamoxifen, a selective ER modulator, is the most widely used anti-estrogen therapy for premenopausal and postmenopausal women with metastatic breast cancer, for adjuvant treatment of primary breast cancer, and as a chemopreventative agent for women with a high risk of developing breast cancer [6,11]. Tamoxifen needs to be converted to the metabolite endoxifen for biological activity, and this conversion is catalyzed by the enzyme cytochrome P450 2D6 (CYP2D6). Polymorphisms in this gene can significantly affect enzymatic activity (Table 2.4), and more than 75 CYP2D6 variant alleles have been reported [82]. Among healthy Europeans, 6–10% are deficient in CYP2D6 metabolism [83]. These individuals convert tamoxifen to endoxifen poorly – a 15-fold lower rate of conversion is observed *in vitro* – and therefore may not derive full therapeutic benefit from tamoxifen therapy. Interestingly, 12 women with breast cancer taking adjuvant tamoxifen were given CYP2D6 inhibitors to assess the role of this enzyme on the levels of endoxifen. Results indicated that the plasma concentrations of endoxifen decreased approximately two-fold after four weeks of exposure to CYP2D6 inhibitors, confirming the role of CYP2D6 in tamoxifen metabolism [84].

A comparison of 1,325 patients after nine years of follow-up revealed that following tamoxifen treatment, patients with the low metabolism genotype (low CYP2D6 activity) have a two-fold increased risk of recurrence compared to those with the high metabolism genotype [85]. Additionally, in a German study of women diagnosed with ER+ primary invasive breast cancer who either received adjuvant tamoxifen monotherapy or did not receive therapy, the tamoxifen-treated patients with intermediate or poor CYP2D6 activity had shorter relapse-free times and event-free survival than patients with normal CYP2D6 metabolism. In patients not treated with tamoxifen, CYP2D6 activity did not affect survival, suggesting that CYP2D6 activity predicts response to tamoxifen, but is not a general prognostic marker for BC patients [86]. Alternatively, other retrospective analyses have not supported a role for CYP2D6 in the metabolism of tamoxifen, showing no association between CYP2D6 genotype and recurrence-free survival in women with ER+ breast cancer treated with tamoxifen [87–89]. Unfortunately, there is a large degree of heterogeneity between the relevant clinical studies, making it difficult to compare them directly to get a better understanding of the discrepancies. What is known, though, is that the studies that did not report an association between CYP2D6 genotype and tamoxifen metabolism also did not measure

TABLE 2.4 The Effect of Multiple CYP2D6 Alleles and their Effect on Enzyme Activity

CYP2D6 Alleles	Allele Designation	Enzyme Activity
*1, *2, *33, *35	Normal or Wild Type	Normal
*3, *4, *5–*8, *11–*16, *18–*21, *36, *38, *40, *42, *44, *56, *62	Null	No protein, inactive or negligible
*9, *10, *17, *29, *41, *59	Reduced Activity	Decreased
*22, *28, *30–*32, *34, *37, *39, *43, *45–*55	Unknown Activity	Unknown

(Adapted from Hoskins et al. [85].)

levels of the active metabolite endoxifen or genotype the rarer CYP2D6 alleles [87–89], making it difficult to fully interpret these results.

In 2006, an FDA advisory panel recommended adding a warning to tamoxifen indicating that CYP2D6 genotyping is an option for women with BC being considered for this treatment, although the ASCO guidelines recommended against it at the time [90]. Furthermore, multiple retrospective analyses support the use of CYP2D6 genotype to guide the dose and selection of tamoxifen, but this testing has yet to be made mandatory. If the predictive ability of CYP2D6 genotyping results is confirmed in a prospective trial, this evaluation may become routine for ER+ patients in order to identify those who should consider alternative therapies to block activation of the ER, such as aromatase inhibitors.

2.2.3 Prognostic Biomarkers

Prognostic biomarkers differentiate between cancer patients likely to have different outcomes and may determine what therapy is utilized, as well as define the course of treatment. General examples of prognostic biomarkers utilized across multiple cancer indications include cancer stage, tumor grade, and existence of distant metastases. A more specific example exists in CRC where stage, grade, and extent of vascular invasion are used in combination to describe a subgroup of stage II colon cancer patients who may have improved potential for survival from adjuvant chemotherapy [91]. Recently, molecularly-based tests have emerged as prognostic biomarkers, particularly for breast cancer patients [92,93], but for other indications as well.

Oncotype DX (Genomic Health) is a 21-gene transcript-based assay indicating the probability of BC recurring after surgical intervention [94–97]. This test was developed specifically as a prognostic test to assess the benefit of chemotherapy in women with node-negative, ER+ breast cancer who have been treated with tamoxifen, as well as identifying patients at greatest risk of relapse following treatment [94–97]. To develop this test, a real-time PCR method was used in FFPE tumor samples to quantify expression of 250 candidate genes selected from the literature based on data from fresh-frozen tissue [98]. Three independent studies comprising a total of 447 patients were used to evaluate the relationship between the candidate genes and BC recurrence. Based on these studies the expression of a panel of 16 cancer-related genes and five reference genes was used to calculate a recurrence score for each tumor sample [98]. Figure 2.4

Proliferation	Estrogen	Reference
Ki67	ER	ACTB
STK15	PR	GAPDH
Survivin	BCL2	RPLPO
CCNB1	SCUBE2	GUS
MYBL2		TFRC

Invasion	HER2	Other
MMP11	GRB7	GSTM1
CTSL2	HER2	CD68
		BAG1

FIGURE 2.4   21-gene panel used to calculate recurrence score [98].

shows the genes included in this panel; the algorithm used to calculate a recurrence score from these genes is described in detail by Paik et al. [98].

A recurrence score of <18 is considered low risk, whereas RS ≥18 and <31 is considered intermediate risk, and RS ≥31 is considered high risk for distance recurrence. A randomized trial of 2617 node-negative, hormone receptor positive women taking tamoxifen versus placebo was used to evaluate the 21-gene assay [99]. Of the 668 patients taking tamoxifen, 93% of patients in the low risk group were free from distance recurrence compared to only 69% of the high risk group [98,99]. The probability of developing a distant recurrence was 31% in the high risk group, 14.4% in the intermediate group, and 6.9% in the low risk group. Not only does the recurrence score predict distant recurrence for all age categories and across tumor sizes in multiple trials, but it also predicts overall survival [98]. The 21-gene assay has been endorsed by ASCO as a tumor marker [90] and the National Comprehensive Cancer Network (NCCN) Breast Cancer Panel considers it an option for patients with ER+ node-negative breast cancer for guiding decision making regarding adjuvant chemotherapy. To further increase the ability to interpret Oncotype DX data for therapeutic decision making, the TAILORx study is testing BC patients with the Oncotype DX and then assigning patients with intermediate risk of recurrence to either hormonal therapy alone or chemotherapy and hormonal therapy [100]. This trial will provide key information on this prognostic test regarding the intermediate risk group, where it is currently unclear whether these patients benefit from the addition of chemotherapy [100].

In contrast to Oncotype DX, MammaPrint (Agendia) was developed as a general prognostic test for premenopausal untreated patients, and its role as a predictive marker for a particular treatment response was not examined when initially developed. This test can also be used to identify patients who may benefit from adjuvant chemotherapy + hormone therapy, and who are at greatest risk of relapse following treatment [101,102]. Development of this prognostic test originated with a microarray analysis of the relative expression levels of 25,000 genes in tumors from a cohort of 78 premenopausal patients younger than 55 years without nodal metastasis who received no systemic therapy [103]. By associating the expression of each gene with clinical outcome, a prognostic algorithm based on the expression levels of 70 genes was established. A threshold to separate patients with good and poor prognoses was determined in this discovery cohort of 78 patients, and the average gene expression level of the 70 genes in patients with a good prognosis was calculated [103]. For any new case, a correlation coefficient for the 70 gene expression classifier is developed by comparing the expression levels to the average value for the good prognosis group from the first study cohort. Tumors are classified as having a good prognosis if the correlation coefficient is above 0.4. Turning the 70-gene assay into the MammaPrint assay was achieved by designing a custom microarray with fewer probes than the original array of 25,000 genes, which allows eight patients to be tested on a single microarray slide [104]. As the MammaPrint assay is based on microarray technology, it requires RNA to be obtained from frozen tissue. The inability of this test to be applied to FFPE samples limits its clinical utility.

In a cohort of 80 postmenopausal women with ER+ and node-negative breast cancer, the MammaPrint assay classified 27 patients as low risk and these patients had excellent clinical outcomes [105]. The clinical utility of the MammaPrint assay is currently being tested in a randomized clinical trial called MINDACT (Microarray In Node-Negative and 1 to 3 positive lymph node Disease may Avoid Chemo Therapy), a prospective randomized study comparing the 70-gene signature with common pathological criteria in selecting BC patients with 0–3 positive nodes for adjuvant chemotherapy [106]. Although complex, the study aims to address whether low risk patients defined by MammaPrint have good prognosis when treated with endocrine therapy only, and if patients identified as high risk by pathological parameters but low risk by MammaPrint (about 10–15% of participants) have good prognosis when treated with endocrine therapy only [106]. The I-SPY 2 trial is an additional study examining MammaPrint [107], and this is using an adaptive design to identify improved treatment regimens for patients grouped by particular molecular signatures [107]. In this trial, standard biomarkers such as hormone receptor status, HER2 status, and MammaPrint status are used to assign various treatments and randomize patients to one of five drugs added to standard neoadjuvant chemotherapy [107]. Many of the biomarker signatures evaluated in this trial represent disease areas where there is need for improved treatment, such as hormone receptor and HER2 negative tumors and tumors with poor prognosis based on high MammaPrint scores. Trials utilizing an adaptive design with multiple drugs provide important platforms for the evaluation of numerous biomarker strategies that may be important for the success of new therapeutics.

2.2.4 Resistance Biomarkers

It is often the case that cancer therapies work for a period of time, but then tumors develop resistance to a particular therapy, often in a majority of the patients. Although there are currently no FDA approved resistance biomarkers for use with specific drugs, many are being investigated and showing promise. For EGFR TKIs, one resistance mechanism involves secondary mutations in EGFR, where the best studied is EGFR T790M in exon 20, found in approximately 50% of patients with acquired resistance [108–110]. This mutation in EGFR can also be detected in plasma, making it possible that circulating EGFR T790M could be used in the early detection of resistance [111] and to identify patients in need of alternative therapeutics.

Resistance to imatinib occurs in about 10–15% of CML patients, often due to mutations in the catalytic domain of the BCR-ABL fusion protein [112]. With this knowledge of resistance mechanisms to imatinib, other drugs have been identified to work in imatinib-resistant CML, such as dasatinib and nilotinib [113–116]. Both of these TKIs have activity against other kinases, which may help explain their efficacy in this setting [113–116]. Molecular analyses have been utilized to screen for catalytic domain mutations in BCR-ABL that might play a role in resistance, which could ultimately be used to select alternative treatments to imatinib for patients likely to develop resistance.

2.2.5 Surrogate Biomarkers

A biomarker can also serve as a surrogate endpoint when it is used to substitute for an established clinical endpoint. To be a direct substitute for a clinical benefit endpoint, a surrogate endpoint must be correlated with the clinical outcome and must fully capture the net effect of treatment on the clinical outcome. One example of this principle is that a drug that lowers low density lipoprotein (LDL) can be approved by the FDA on that basis alone without the requirement for direct clinical evidence of the reduction of risk of cardiovascular disease. The benefit of reduced LDL on cardiovascular risk is so well established that reduction in the surrogate marker is sufficient for approval. In the oncology field, the lack of such well-established surrogate biomarkers of anti-tumor activity or overall survival has hampered early clinical development of new therapies/therapeutic combinations.

The enumeration and characterization of circulating tumor cells (CTCs) prior to and during various treatment regimens have the potential to become a surrogate endpoint that could accelerate development of new anti-tumor agents. CTCs are present in the blood of many patients with cancer and can be detected/quantified using the FDA approved CellSearch system from Veridex [117]. The number of CTCs per 7.5 mL of blood has been identified as a prognostic biomarker across a range of indications, including breast, colorectal, and prostate cancer. Studies are currently ongoing to measure changes in CTCs following multiple therapies both to evaluate its utility as a predictive biomarker and evaluate its association with established clinical endpoints, such as disease-free survival and overall survival. Despite considerable effort in this area, the use of CTC numbers as a surrogate biomarker/surrogate endpoint for antitumor activity or as a guide for therapeutic intervention is not yet part of routine clinical practice [118]. Limitations currently exist in the detection of CTC, in that the use of anti-EpCAM-coated beads to identify circulating epithelial cells from various tumors does not work when EpCAM expression levels are low [118]. Furthermore, the need for fresh blood samples and the inability to both enumerate CTCs and evaluate their molecular phenotype in a clinical setting add to the challenges of utilizing CTCs as surrogate biomarkers. Nonetheless, as technology improves to overcome these current limitations, the association of CTCs with clinical benefit and utility as a surrogate biomarker may be possible [117–119]. Having such a surrogate endpoint will aid clinical decision making and positively impact drug development timelines in early phase trials where it is challenging to assess overall survival due to lengthy timelines.

2.3 COMPANION DIAGNOSTICS

Most of the biomarkers that have successfully reached the clinic have been identified though retrospective analyses. Existing knowledge of mutations and other molecular data available early in the drug development process has yet to become a routine aspect of prospective trials. Thus, the challenge is to use current techniques and knowledge to pursue opportunities for predictive biomarkers to be identified prospectively or co-developed as companion diagnostics.

Companion diagnostics are *in vitro* diagnostic (IVD) tests that are intended to insure that therapeutic products are used according to their label and are safe and effective [120–122]. Additionally, they are of substantial importance in patient selection for certain treatments and in preventing undue safety risk that may arise from patients receiving treatments unlikely to work. The term 'companion diagnostic' can be used independently of where in the drug development cycle the diagnostic test is introduced. For example, a companion diagnostic can be developed with a new drug, as in the case of HER2 for trastuzumab/Herceptin [18]. A new diagnostic test can be developed for an existing drug, as in the case of UGT1A1 for the chemotherapeutic irinotecan [73,75,76]. Finally, an existing diagnostic test could be applied for a new drug. Since 1994, approximately 100 molecular IVD tests have received pre-market authorization from the FDA, with about a quarter of these tests relying on nucleic acid detection [123,124]. The majority of IVDs used for oncology indications are for breast cancer (36%) and prostate cancer (31%), and oncology IVDs are the only ones currently approved to yield data supporting specific treatment selection [123–125]. Companion diagnostics are likely to become increasingly important as researchers continue to dissect particular patient subsets within various cancer indications that respond to specific therapies. Additionally, it has become clear that a validated companion diagnostic provides tremendous support and can often accelerate FDA approval of new therapeutics.

Successful recent examples of companion diagnostics utilized in oncology include those that accompany the FDA approval of two relatively new therapies, crizotinib (Pfizer) and vemurafenib (Roche) in advanced or metastatic NSCLC and metastatic melanoma, respectively. Approvals for both drugs were dependent on simultaneous companion diagnostic test approvals – the Vysis ALK Break-Apart FISH Probe Kit (Abbott Molecular, Inc.) with crizotinib [55,56,61] and the cobas 4800 BRAF V600 Mutation Test (Roche Molecular Systems) with vemurafenib [66,67]. Crizotinib received accelerated FDA approval based on Phase II data due to the existence of a prospective genomic biomarker and parallel companion diagnostic development. It is the first drug approval that occurred within four years of identifying its role in disease. These new approvals join the growing list of other drug/diagnostic combinations used in clinical practice, such as anti-EGFR antibodies/K-ras mutations, gefitinib/EGFR mutations, trastuzumab/HER2 levels and tamoxifen/ER expression. Approvals of crizotinib and vemurafenib may represent a new paradigm for the next generation of targeted therapies in oncology, in that large changes in patient outcomes are possible when molecularly defined sub-populations are identified appropriately. Additionally, the development of both drugs illustrates the benefit of incorporating hypothesis testing into early clinical trials when possible because it permits the collection of appropriate samples to evaluate multiple biomarker strategies.

The concept of testing multiple biomarker strategies is beginning to be incorporated creatively into clinical trial design to address the absolute need for predictive biomarkers to guide treatment selection for patients with various diseases. One specific example where this concept is particularly relevant is in NSCLC, where only two drugs that target signaling pathways have been approved by the FDA in unselected NSCLC patients: erlotinib (EFGR TKI) and bevacizumab (mAb targeting vascular endothelial growth factor). In contrast, nine drugs in 13 Phase III trials of unselected patients with NSCLC have failed recently [126,127], due, in part, to a lack of predictive markers. To address this, a novel Phase II Biomarker-integrated Approaches of Targeted Therapy for Lung Cancer Elimination (BATTLE) trial was designed, in which NSCLC tumors are prospectively biopsied and assigned to the treatment with greatest potential benefit using adaptive randomization (see Fig. 2.5 for overall trial schema). Agents were selected due to high scientific and clinical interest at the time (2005) and included EGFR (erlotinib), KRAS/BRAF (sorafenib), retinoid-EGFR signaling (bexarotene and erlotinib), and vascular endothelial growth factor receptor (vandetanib) [126]. Patients were enrolled in the BATTLE study with equally random assignments for the first 97 patients and adaptive randomization for the remaining 158. Biomarker analysis was done in real time, with a large panel of mutation, gene copy number, and IHC analyses performed on each sample. Patients were grouped into predefined biomarker signature groups, which were used to evaluate treatment-biomarker interactions. After equal randomization of the initial 97 patients, the eight week disease control rate (DCR) was evaluated and compared with the biomarker status

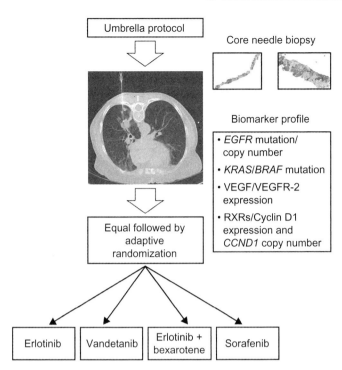

FIGURE 2.5 Schema for the BATTLE study. *(Reprinted from Kim et al. [126].)*

within each treatment arm. Using the biomarker-DCR results, future randomization probabilities were adjusted using a Bayesian model, such that if a patient presented with a particular biomarker signature, he or she had a >25% chance of being randomized to a treatment on which prior patients with the same biomarker signature had positive results.

Results from the BATTLE study revealed interesting biomarker-treatment relationships that had previously not been extensively explored. For example, a better eight week DCR was observed for patients with EGFR amplification receiving erlotinib plus bexarotene; a worse eight week DCR was observed for patients with EGFR mutations receiving sorafenib; and, compared to other treatments tested, sorafenib had a higher DCR (64% versus 33%) in EGFR-wild-type patients and a nonstatistically significant trend toward better DCR (61% versus 32%) in KRAS-mutant patients. Another key finding was a promising relationship between KRAS or BRAF mutation-positive patients and sorafenib therapy. Biomarker-positive patients had a 79% eight week DCR compared to a 14% eight week DCR with erlotinib [126–128]. The adaptive randomization design likely enhanced the ability to make this observation by preferentially randomizing KRAS or BRAF mutation-positive patients to the sorafenib arm. The success of the BATTLE trial in demonstrating the feasibility of the novel design and potential to reveal informative biomarker associations has already inspired similar designs, such as the BATTLE-2 trial, which will further refine the approach taken in the BATTLE study by pre-specifying a limited set of markers in the first half of the study population (200 patients) to conduct prospective testing of biomarkers/signatures. The most predictive markers and signatures will then be used to guide patient assignments to the most favorable matched treatments in the second half of the study (200 patients). In addition to the BATTLE-2 study, the Lung Cancer Mutation Consortium (LCMC), an initiative of the National Cancer Institute comprised of 14 leading cancer centers across the US, has formed to evaluate the frequency of key mutations in NSCLC, including KRAS, EGFR, HER2, BRAF, PIK3CA, AKT1, MEK1, and NRAS using standard multiplexed assays and ALK rearrangements and MET amplifications using FISH [129]. In 2011, 830 patients had been registered and 60% of these tumors had a driver mutation detected, 95% of which were mutually exclusive [129]. Patients with specific mutations are offered participation in LCMC-linked trials where their mutational status would suggest a higher probability of success. Creative trials and multi-center initiatives with the goal of evaluating key biomarker-treatment relationships will continue to advance the field and identify new markers that could become diagnostic tests.

The application of molecular diagnostics will continue to revolutionize the drug discovery and development process, with gene-based and molecular diagnostics testing growing at a 30–50% rate [120,123,124]. As many as 1500 genes and 5000 proteins are currently candidates for new molecular tests, and diagnostic divisions already exist in many companies to combine drug development with the production of diagnostic tests. Currently, the use of companion diagnostics for patient screening in clinical trials is surprisingly infrequent, often due to the challenges in identifying the exact genetic cause for most diseases where this remains unknown, and the likelihood that multiple factors are important for disease development/progression, making it difficult to identify a test that would yield a definitive response. However, the molecular diagnostics industry is constantly evolving. New technologies such as next generation sequencing (NGS) and new disease associations/clinical opportunities for predicting efficacy and monitoring disease outcome are constantly emerging.

Standardization of NGS workflows could substantially reduce the cost of companion diagnostic assay development, as most of the validation around instrumentation and sequencing protocols would already be in place; therefore, a new assay would only need to be validated with respect to the genomic value in a particular cancer type or patient subset. Furthermore, NGS could have a central role in the discovery of new genomic biomarkers since many different types of experiments can be performed on a single machine. For example, the FDA recognizes 41 genes or chromosomal rearrangements as pharmacogenomic biomarkers, most of which are evaluated by laboratory tests [130], but NGS could allow all biomarkers to be combined (plus some others) in a single test. In support of this concept, a panel that sequences 92% of all genomic rearrangements relevant to TKIs has been developed recently [131]. For known alterations, it would be more cost-effective to use these types of NGS gene panels instead of multiple individual diagnostic tests, so as we increase our understanding of various cancers, and as new drugs become available, these types of panels may make the development of companion diagnostics more feasible.

2.4 CONCLUSIONS

Clinical innovation always seems to take longer than expected. A decade after completion of the human genome sequence we still only have a handful of FDA approved companion diagnostics that predict response to therapy. These successes point to a promising future where we will be able to utilize targeted therapies much more effectively in cancer patients, and will be able to identify mechanisms of drug resistance and areas of unmet need much more quickly. However, it is sobering to realize how difficult it has been to develop biomarkers that actually impact clinical practice. Currently, most of our successful examples in oncology are limited to a few drugs inhibiting signal transduction pathways. These drugs have become models for biomarker strategies and have several unique advantages that are not always available for other classes of drug. So, what makes signal transduction pathways good opportunities for biomarker development? Signal transduction pathways are often activated by 'driver' mutations that can be easily detected by sequencing, PCR or FISH. Consequently, it is possible to surmise that those tumor cells with the driver mutation may be susceptible to a therapeutic intervention that blocks signaling through this pathway identified by the cognate mutation. For example, the 50% of melanomas with BRAF signaling activated by the V600E mutation will respond to inhibition of BRAF by vemurafenib, while this drug will have no effect on melanomas without BRAF activation. So, pathway activation biomarkers have two enormous advantages for the development of predictive biomarkers. First, the markers are easy to identify by scanning for recurrent mutations at the target gene or downstream signal transduction genes, and, secondly, there is an enormous difference in effect size between the subgroups which enables the development of highly predictive markers to differentiate the clinical response between the marker positive and negative classes.

The development of molecular profiles has proved particularly challenging for oncology drugs. Indeed, the only FDA approved molecular profiles are for prognostic markers and not predictive biomarkers. These prognostic tests (MammaPrint – Agendia, Tumor of Origin – Pathwork Diagnostics, and Allomap – XDx) were all developed against large retrospective databases with known disease outcome or tumor of origin; data sets that do not exist for first-in-class drugs during development. A further complexity for the development of oncology drugs is that the endpoint changes during development. Early clinical studies are done using response (or tumor shrinkage) as the primary endpoint. Late development and registrational trials are done using progression-free survival (PFS) or overall survival (OS) as endpoints. Unfortunately, predictive markers for response developed in the early clinical trials have often not predicted PFS or OS accurately.

Accelerating personalized healthcare for cancer patients will require thoughtful implementation of new technologies (notably NGS) in larger, innovative clinical trials. We are entering a positive reinforcement cycle where new targeted therapies drive the need for more predictive biomarkers and more predictive biomarkers create niches for new targeted therapies. Today, we have 12 targeted drugs in approved in oncology against only receptor tyrosine kinase or signal transduction targets. Ongoing Phase IIB and III clinical trials will likely lead to the approval of many more targeted therapies in the next few years, resulting in the need to profile all cancer patients to determine eligibility for treatment with a targeted therapy. Once on treatment with a targeted therapy, it will become important to use serum DNA and CTCs to monitor treatment response and the development of acquired drug resistance and to alter therapy accordingly. Cancer gene panel, exome, and eventually whole genome sequencing will soon be routinely applied to all metastatic cancer patients, and we will be able to determine the success of personalized healthcare strategies in oncology by demonstrating increased PFS and OS for patients being treated with targeted therapies guided in real time by predictive biomarkers.

References

[1] Walker I, Newell H. Do molecularly targeted agents in oncology have reduced attrition rates? Nat Rev Drug Discov 2009;8(1):15–16.

[2] Deininger MW, Druker BJ. Specific targeted therapy of chronic myelogenous leukemia with imatinib. Pharmacol Rev 2003;55:401–23.

[3] Wodarz D. Heterogeneity in chronic myeloid leukaemia dynamics during imatinib treatment: role of immune responses. Proc Biol Sci 2010;277:1875–80.

[4] Beatson GW. On the treatment of inoperable cases of carcinoma of the mamma: suggestions of a new method of treatment with illustrative cases. Lancet 1896;2:162–5.

[5] De Sombre ER, Smith S, Block GE, Ferguson DJ, Jensen EV. Prediction of breast cancer response to endocrine therapy. Cancer Chemother Rep 1974;58:513–19.

[6] Puhalla S, Bhattacharya S, Davidson N. Hormonal therapy in breast cancer: a model disease for the personalization of cancer care. Molec Oncol 2012;6:222–36.

[7] Allred DC, Carlson RW, Berry DA, Burstein HJ, Edge SB, Goldstein LJ, et al. NCCN Task force report: estrogen receptor and progesterone receptor testing in breast cancer by immunohistochemistry. J Natl Compr Canc Netw 2009;7(Suppl. 6):S1–S21.

[8] Harvey JM, Clark GM, Osborne CK, Allred DC. Estrogen receptor status by immunohistochemistry is superior to the ligand-binding assay for predicting response to adjuvant endocrine therapy in breast cancer. J Clin Oncol 1999;17:1474–81.

[9] McGuire WL, Carbone PP, Sears ME, Escher GC. Estrogen receptors in human breast cancer: an overview. In: McGuire WL, Carbone PP, Vollner EP, editors. Estrogen receptors in human breast cancer. New York: Raven Press; 1975. pp. 1–8.

[10] Early Breast Cancer Trialist's Collaborative Group Tamoxifen for early breast cancer: an overview of randomized trials. Lancet 1998;351:1451–67.

[11] Early Breast Cancer Trialists' Collaborative Group (EBCTCG) Effect of chemotherapy and hormone therapy for early breast cancer on recurrence and 15-year survival: an overview of randomised trials. Lancet 2005;365:1687–717.

[12] Duffy MJ. Estrogen receptors: role in breast cancer. Crit Rev Clin Lab Sci 2006;43:325–47.

[13] Musgrove EA, Sutherland RL. Biological determinants of endocrine resistance in breast cancer. Nat Rev Cancer 2009;9:631–43.

[14] Slamon DJ, Clark GM, Wong SG, Levin WJ, Ullrich A, McGuire WL. Human breast cancer: correlation of relapse and survival with amplification of the HER-2/neu oncogene. Science 1987;235:177–82.

[15] Browne BC, O'Brien N, Duffy MJ, Crown J, O'Donovan N. HER-2 signaling and inhibition in breast cancer. Curr Cancer Drug Ther 2009;9:419–38.

[16] Ross JS, Slodkowska EA, Symmans WF, Pusztai L, Ravdin PM, Hortobagyi GN. The HER-2 receptor and breast cancer: ten years of targeted anti-HER-2 therapy and personalized medicine. Oncologist 2009;14:320–68.

[17] Dawood S, Broglio K, Buzdar AU, Hortobagyi GN, Giordano SH. Prognosis of women with metastatic breast cancer by HER2 status and trastuzumab treatment: an institutional-based review. J Clin Oncol 2010;28:92–8.

[18] Hudis CA. Trastuzumab, mechanism of action and use in clinical practice. N Engl J Med 2007;357:39–51.

[19] Cobleigh MA, Vogel CL, Tripathy D, Robert NJ, Scholl S, Fehrenbacher L, et al. Multinational study of the efficacy and safety of humanized anti-HER2 monoclonal antibody in women who have HER2 overexpressing metastatic breast cancer that has progressed after chemotherapy for metastatic disease. J Clin Oncol 1999;17:2639–48.

[20] Slamon DJ, Leyland-Jones B, Shak S, Fuchs H, Paton V, Bajamonde A, et al. Use of chemotherapy plus a monoclonal antibody against HER2 for metastatic breast cancer that overexpressed HER2. N Eng J Med 2001;344:783–92.

[21] Dahabreh IJ, Linardou H, Siannis F, Fountzilas G, Murray S. Trastuzumab in the adjuvant treatment of early-stage breast cancer: a systematic review and meta-analysis of randomized controlled trials. Oncologist 2008;13:620–30.

[22] Romond EH, Perez EA, Bryant J, Suman VJ, Geyer CE, Davidson NE. Trastuzumab plus adjuvant chemotherapy for operable HER2-positive breast cancer. N Engl J Med 2005;353:1673–84.

[23] Piccart-Gebhart MJ, Procter M, Leyland-Jones B, Goldhirsch A, Untch M, Smith I. Trastuzumab after adjuvant chemotherapy in HER2-positive breast cancer. N Engl J Med 2005;353:1659–72.

[24] Allison M. The HER2 testing conundrum. Nat Biotechnol 2010;28:117–19.

[25] Dendukuri N, Khetani K, McIsaac M, Brophy J. Testing for HER2-positive breast cancer: a systematic review and cost-effectiveness analysis. CMAJ 2007;176:1429–34.

[26] Perez EA, Baweja M. HER2-positive breast cancer: current treatment strategies. Cancer Invest 2008;26:545–52.

[27] Hicks DG, Kulkarni S. Trastuzumab as adjuvant therapy for early breast cancer: the importance of accurate human epidermal growth factor receptor 2 testing. Arch Pathol Lab Med 2008;132:1008–15.

[28] Valabrega G, Montemurro F, Aglietta M. Trastuzumab: mechanism of action, resistance and future perspectives in HER2-overexpressing breast cancer. Ann Oncol 2007;18:977–84.

[29] Suter TM, Procter M, van Veldhuisen DJ, Muscholl M, Bergh J, Carlomagno C. Trastuzumab-associated cardiac adverse effects in the herceptin adjuvant trial. J Clin Oncol 2007;25:3859–65.

[30] Topalian SL, Drake CG, Pardoll DM. Targeting the PD-1/B7-H1(PD-L1) pathway to activate anti-tumor immunity. Curr Opin Immunol 2012;24(2):207–12.

[31] Topalian SL, Hodi S, Brahmer JR, Gettinger SN, Smith DC, McDermott DF, et al. Safety, activity, and immune correlates of anti–PD-1 antibody in cancer. N Engl J Med 2012;366:2443–54.

[32] Ciardiello F, Tortora G. EGFR antagonists in cancer treatment. N Engl J Med 2008;358:1160–74.

[33] Dahabreh IJ, Linardou H, Siannis F, Kosmidis P, Bafaloukos D, Murray S. Somatic EGFR mutation and gene copy gain as predictive biomarkers for response to tyrosine kinase inhibitors in non-small cell lung cancer. Clin Cancer Res 2010;16:291–303.

[34] Saif MW. Colorectal cancer in review: the role of the EGFR pathway. Expert Opin Investig Drugs 2010;19: 357–69.

[35] Linardou I, Dahabreh D, Kanaloupiti F, Siannis D, Bafaloukos P, Kosmidis C, et al. Assessment of somatic k-RAS mutations as a mechanism associated with resistance to EGFR-targeted agents: a systematic review and meta-analysis of studies in advanced non-small-cell lung cancer and metastatic colorectal cancer. Lancet Oncol 2008;9:962–72.

[36] Maemondo M, Inoue A, Kobayashi K, Sugawara S, Oizumi S, Isobe H. Gefitinib or chemotherapy for non-small-cell lung cancer with mutated EGFR. N Engl J Med 2010;362:2380–8.

[37] Saijo N. Targeted therapies: tyrosine-kinase inhibitors – a new standard for NSCLC therapy. Nat Rev Clin Oncol 2010;7:618–19.

[38] Santini D. Molecular predictive factors of response to anti-EGFR antibodies in colorectal cancer patients. Eur J Cancer Suppl 2008;6:86–90.

[39] Karapetis CS, Khambata-Ford S, Jonker DJ, O'Callaghan CJ, Tu D, Tebbutt NC, et al. K-ras mutations and benefit from cetuximab in advanced colorectal cancer. N Engl J Med 2008;359:1757–65.

[40] Mack GS. FDA holds court on post hoc data linking KRAS status to drug response. Nat Biotechnol 2009;27:110–12.

[41] Massarelli E, Varella-Garcia M, Tang X, Xavier AC, Ozburn NC, Liu DD, et al. KRAS mutation is an important predictor of resistance to therapy with epidermal growth factor receptor tyrosine kinase inhibitors in non-small-cell lung cancer. Clin Cancer Res 2007;13:2890–6.

[42] Shankaran V, Obel J, Benson III AB. Predicting response to EGFR inhibitors in metastatic colorectal cancer: current practice and future directions. Oncologist 2010;15:157–67.

[43] Lopez-Chavez A, Giaccone G. Targeted therapies: importance of patient selection for EGFR TKIs in lung cancer. Nat Rev Clin Oncol 2010;7:360–2.

[44] Parra HS, Cavina R, Latteri F, Zucali PA, Campagnoli E, Morenghi E, et al. Analysis of epidermal growth factor receptor expression as a predictive factor for response to gefitinib ('Iressa', ZD1839) in non-small-cell lung cancer. Br J Cancer 2004;19:208–12.

[45] Perez-Soler R, Chachoua A, Hammond LA, Rowinsky EK, Huberman M, Karp D. Determinants of tumor response and survival with erlotinib in patients with non-small-cell lung cancer. J Clin Oncol 2004;22:3238–47.

[46] Douillard JY, Shepherd FA, Hirsh V, Mok T, Socinski MA, Gervais R, et al. Molecular predictors of outcome with gefitinib and docetaxel in previously treated non-small cell lung cancer: data from the randomized phase III INTEREST trial. J Clin Oncol 2010;28:744–52.

[47] Mitsudomi T, Morita S, Yatabe Y, Negoro S, Okamoto I, Tsurutani J, et al. Gefitinib versus cisplatin plus docetaxel in patients with non-small-cell lung cancer harboring mutations of the epidermal growth factor receptor (WJTOG3405): an open label, randomized Phase III trial. Lancet Oncol 2010;11:121–8.

[48] Hammerman PS, Janne PA, Johnson BE. Resistance to epidermal growth factor receptor tyrosine kinase inhibitors in non-small cell lung cancer. Clin Cancer Res 2009;15:7502–9.

[49] Mok T, Wu YL, Thongprasert S, Yang CH, Chu DT, Saijo N, et al. Gefitinib or carboplatin-paclitaxel in pulmonary adenocarcinoma. N Engl J Med 2009;361:947–57.

[50] D'Addario G, Felip E, On behalf of the ESMO Guidelines Working Group Nonsmall cell lung cancer: ESMO clinical recommendations for diagnosis, treatment and follow-up. Ann Oncol 2008;19(Suppl. 2):ii39–40.

[51] Jabbour E, Kantarjian H. Chronic myeloid leukemia: 2012 Update on diagnosis, monitoring, and management. Am J Hematol 2012;87:1038–45.

[52] Deininger M, O'Brien SG, Guilhot F, et al. International randomized study of interferon vs. STI571 (IRIS) 8-year follow up: Sustained survival and low risk for progression of events in patients with newly diagnosed chronic myeloid leukemia in chronic phase (CML-CP) treated with imatinib. Blood 2009;114:1126.

[53] Druker BJ, Talpaz M, Resta DJ, Peng B, Buchdunger E, Ford JM, et al. Efficacy and safety of a specific inhibitor of the BCR-ABL tyrosine kinase in chronic myeloid leukemia. N Engl J Med 2001;344(14):1031–7.

[54] Kwak EL, Bang YJ, Camidge DR, Shaw AT, Solomon B, Maki RG, et al. Anaplastic lymphoma kinase inhibition in non-small-cell lung cancer. N Engl J Med 2010;363(18):1693–703.

[55] Shaw AT, Yeap BY, Solomon BJ, Riely GJ, Gainor J, Engelman JA, et al. Effect of crizotinib on overall survival in patients with advanced non-small-cell lung cancer harboring ALK gene rearrangement: a retrospective analysis. Lancet Oncol 2011;12(11):1004–12.

[56] Ou SH, Bazhenova L, Camidge DR, Solomon BJ, Herman J, Kain T, et al. Rapid and dramatic radiographic and clinical response to an ALK inhibitor (crizotinib, PF02341066) in an ALK translocation-positive patient with non-small cell lung cancer. J Thorac Oncol 2010;5(12):2044–6.

[57] Forde PM, Rudin CM. Crizotinib in the treatment of non-small-cell lung cancer. Expert Opin Pharmacother 2012;13(8):1195–201.

[58] Shaw AT, Yeap BY, Mino-Kenudson M, et al. Clinical features and outcome of patients with non-small-cell lung cancer who harbor EML4–ALK. J Clin Oncol 2009;27(26):4247–53.

[59] Choi YL, Takeuchi K, Soda M, Inamura K, Togashi Y, Hatano S, et al. Identification of novel isoforms of the EML4–ALK transforming gene in non-small cell lung cancer. Cancer Res 2008;68(13):4971–6.

[60] Wong DW, Leung EL, So KK, Tam IY, Sihoe AD, Cheng LC, et al. The EML4-ALK fusion gene is involved in various histologic types of lung cancers from nonsmokers with wild-type EGFR and KRAS. Cancer 2009;115(8):1723–33.

[61] Ou SH, Bartlett CH, Mino-Kenudson M, Cui J, Iafrate AJ. Crizotinib for the treatment of ALK-rearranged non-small cell lung cancer: a success story to usher in the second decade of molecular targeted therapy in oncology. Oncologist 2012;17(11):1351–75.

[62] Diverio D, Riccioni R, Mandelli F, Lo Coco F. The PML/RAR alpha fusion gene in the diagnosis and monitoring of acute promyelocytic leukemia. Haematologica 1995;80:155–60.

[63] Lin RJ, Evans RM. Acquisition of oncogenic potential by RAR chimeras in acute promyelocytic leukemia through formation of homodimers. Mol Cell 2000;5:821–30.

[64] Zhou DC, Kim SH, Ding W, Schultz C, Warrell Jr RP, Gallagher RE. Frequent mutations in the ligand-binding domain of PML-RARalpha after multiple relapses of acute promyelocytic leukemia: analysis for functional relationship to response to all-trans retinoic acid and histone deacetylase inhibitors *in vitro* and *in vivo*. Blood 2002;99:1356–63.

[65] Tang XH, Gudas LJ. Retinoids, retinoic acid receptors, and cancer. Annu Rev Pathol 2011;6:345–64.

[66] Flaherty KT, Puzanov I, Kim KB, Ribas A, McArthur GA, Sosman JA, et al. Inhibition of mutated, activated BRAF in metastatic melanoma. N Engl J Med 2010;363:809–19.

[67] Bollag G, Tsai J, Zhang J, Zhang C, Ibrahim P, Nolop K, et al. Vemurafenib: the first drug approved for BRAF-mutant cancer. Nat Rev Drug Discov 2012;11(11):873–86.

[68] Pollock PM, Meltzer PS. A genome-based strategy uncovers frequent *BRAF* mutations in melanoma. Cancer Cell 2002;2:5–7.

[69] Halait H, Demartin K, Shah S, Soviero S, Langland R, Cheng S, et al. Analytical performance of a real-time PCR-based assay for V600 mutations in the BRAF gene, used as the companion diagnostic test for the novel BRAF inhibitor vemurafenib in metastatic melanoma. Diagn Mol Pathol 2012;21:1–8.

[70] Sosman JA, Kim KB, Schuchter L, Gonzalez R, Pavlick AC, Weber JS, et al. Survival in BRAF V600-mutant advanced melanoma treated with vemurafenib. N Engl J Med 2012;366:707–14.

[71] Chapman PB, Hauschild A, Robert C, Haanen JB, Ascierto P, Larkin J, et al. Improved survival with vemurafenib in melanoma with BRAF V600E mutation. N Engl J Med 2011;364:2507–16.

[72] Lopez-Rios F, Angulo B, Gomez B, Mair D, Martinez R, Conde E, et al. Comparison of testing methods for the detection of BRAF V600E mutations in malignant melanoma: Pre-approval validation study of the companion diagnostic test for vemurafenib. PLoS ONE 2013;8(1):e53733.

[73] Hoskins JM, McLeod HL. UGT1A and irinotecan toxicity: keeping it in the family. J Clin Oncol 2009;27(15):15 2419–21.

[74] Satoh T, Ura T, Yamada Y, Yamazaki K, Tsujinaka T, Munakata M, et al. Genotype-directed, dose-finding study of irinotecan in cancer patients with UGT1A1*28 and/orUGT1A1*6 polymorphisms. Cancer Sci 2011;102(10):1868–73.

[75] Marques SC, Ikediobi ON. The clinical application of UGT1A1 pharmacogenetic testing: gene-environment interactions. Hum Genomics 2010;4(4):238–49.

[76] Perera MA, Innocenti F, Ratain MJ. Pharmacogenetic testing for uridine diphosphate glucuronosyltransferase 1A1 polymorphisms: are we there yet? Pharmacotherapy 2008;28:755–68.

[77] Zhang D, Zhang D, Cui D, Gambardella J, Ma L, Barros A, et al. Characterization of the UDP glucuronosyl-transferase activity of human liver microsomes genotyped for the UGT1A1*28 polymorphism. Drug Metab Dispos 2007;35:2270–80.

[78] Stocco G, Crews KR, Evans. WE. Genetic polymorphism of inosinetriphosphate-pyrophosphatase influences mercaptopurine metabolism and toxicity during treatment of acute lymphoblastic leukemia individualized for thiopurine-S-methyl-transferase status. Expert Opin Drug Saf 2010;9(1):23–37.

[79] Relling MV, Hancock ML, Rivera GK, Sandlund JT, Ribeiro RG, Krynetski EY, et al. Mercaptopurine therapy intolerance and heterozygosity at the thiopurine S-methyltransferase gene locus. J Natl Cancer Inst 1999;91(23):2001–8.

[80] Yates CR, Krynetski EY, Loennechen T, Fessing MY, Tai HL, Pui CH, et al. Molecular diagnosis of thiopurine S-methyltransferase deficiency: genetic basis for azathioprine and mercaptopurine intolerance. Ann Intern Med 1997;126(8):608–14.

[81] Stocco G, Cheok MH, Crews KR, Dervieux T, French D, Pei D, et al. Genetic polymorphism of inosine triphosphate pyrophosphatase is a determinant of mercaptopurine metabolism and toxicity during treatment for acute lymphoblastic leukemia. Clin Pharmacol Ther 2009;85(2):164–72.

[82] Ingelman-Sundberg M, Sim SC, Gomez A, Rodriguez-Antona C. Influence of cytochrome P450 polymorphisms on drug therapies: pharmacogenetic, pharmacoepigenetic and clinical aspects. Pharmacol Ther 2007;116:496–526.

[83] Ingelman-Sundberg M. Genetic polymorphisms of cytochrome P450 2d6 (*CYP2D6*): clinical consequences, evolutionary aspects and functional diversity. Pharmacogenomics J 2005;5:6–13.

[84] Hemeryck A, Belpaire FM. Selective serotonin reuptake inhibitors and cytochrome p-450 mediated drug-drug interactions: an update. Curr Drug Metab 2002;3:13–37.

[85] Hoskins JM, Carey LA, McLeod HL. CYP2D6 and tamoxifen: DNA matters in breast cancer. Nat Rev Cancer 2009;9:576–86.

[86] Schroth W, Antoniadou L, Fritz P, Schwab M, Muerdter T, Zanger UM, et al. Breast cancer treatment outcome with adjuvant tamoxifen relative to patient CYP2D6 and CYP2C19 genotypes. J Clin Oncol 2007;25:5187–93.

[87] Nowell SA, Ahn J, Rae JM, Scheys JO, Trovato A, Sweeney C, et al. Association of genetic variation in tamoxifen-metabolizing enzymes with overall survival and recurrence of disease in breast cancer patients. Breast Cancer Res Treat 2005;91:249–58.

[88] Wegman P, Vainikka L, Stål O, Nordenskjöld B, Skoog L, Rutqvist LE, et al. Genotype of metabolic enzymes and the benefit of tamoxifen in postmenopausal breast cancer patients. Breast Cancer Res 2005;7:R284–90.

[89] Wegman P, Elingarami S, Carstensen J, Stål O, Nordenskjöld B, Wingren S. Genetic variants of CYP3A5, CYP2D6, SULT1A1, UGT2B15 and tamoxifen response in postmenopausal patients with breast cancer. Breast Cancer Res 2007;9:R7.

[90] Harris L, Fritsche H, Mennel R, Norton L, Ravdin P, Taube S, et al. update of recommendations for the use of tumor markers in breast cancer. J Clin Oncol 2007;25:5287–312.

[91] Mroczkowski P, Schmidt U, Sahm M, Gastinger I, Lippert H, Kube R. Prognostic factors assessed for 15,096 patients with colon cancer in stages I and II. World J Surg 2012;36(7):1693–8.

[92] Albain KS, Paik S, van't Veer L. Prediction of adjuvant chemotherapy benefit in endocrine responsive, early breast cancer using multigene assays. Breast 2009;18(Suppl. 3):S141–5.

[93] Sotiriou C, Pusztai L. Gene-expression signatures in breast cancer. N Engl J Med 2009;360:790–800.

[94] Paik S, Tang G, Shak S, Kim C, Baker J, Kim W, et al. Gene expression and benefit of chemotherapy in women with node-negative, ER-positive breast cancer. J Clin Oncol 2006;24:3726–34.

[95] Albain KS, Barlow WE, Shak S, Hortobagyi GN, Livingston RB, Yeh IT, et al. Prognostic and predictive value of the 21-gene recurrence score assay in postmenopausal women with node-positive, oestrogen-receptor-positive breast cancer on chemotherapy: a retrospective analysis of a randomised trial. Lancet Oncol 2010;11:55–65.

[96] Joh JE, Esposito NN, Kiluk JV, Laronga C, Lee MC, Loftus L, et al. The effect of Oncotype DX recurrence score on treatment recommendations for patients with estrogen receptor-positive early stage breast cancer and correlation with estimation of recurrence risk by breast cancer specialists. Oncologist 2011;16(11):1520–6.

[97] Kelly CM, Krishnamurthy S, Bianchini G, Litton JK, Gonzalez-Angulo AM, Hortobagyi GN, et al. Utility of oncotype DX risk estimates in clinically intermediate risk hormone receptor-positive, HER2-normal, grade II, lymph node-negative breast cancers. Cancer 2010;116(22):5161–7.

[98] Paik S, Shak S, Tang G, Kim C, Baker J, Cronin M, et al. A multigene assay to predict recurrence of tamoxifen-treated, node-negative breast cancer. N Engl J Med 2004;351:2817–26.

[99] Fisher B, Costantino J, Redmond C, Poisson R, Bowman D, Couture J, et al. A randomized clinical trial evaluating tamoxifen in the treatment of patients with node-negative breast cancer who have estrogen-receptor-positive tumors. N Engl J Med 1989;320:479–84.

[100] Sparano JA. TAILORx: trial assigning individualized options for treatment (Rx). Clin Breast Cancer 2006;7:347–50.

[101] Straver ME, Glas AM, Hannemann J, Wesseling J, van de Vijver MJ, Rutgers EJ, et al. The 70-gene signature as a response predictor for neoadjuvant chemotherapy in breast cancer. Breast Cancer Res Treat 2010;119:551–8.

[102] Knauer M, Mook S, Rutgers EJ, Bender RA, Hauptmann M, van de Vijver MJ, et al. The predictive value of the 70-gene signature for adjuvant chemotherapy in early breast cancer. Breast Cancer Res Treat 2010;120:655–61.

[103] van 't Veer LJ, Dai H, van de Vijver MJ, He YD, Hart AA, Mao M, et al. Gene expression profiling predicts clinical outcome of breast cancer. Nature 2002;415:530–6.

[104] Glas AM, Floore A, Delahaye LJ, Witteveen AT, Pover RC, Bakx N, et al. Converting a breast cancer microarray signature into a high-throughput diagnostic test. BMC Genomics 2006;7:278.

[105] Wittner BS, Sgroi DC, Ryan PD, Bruinsma TJ, Glas AM, Male A, et al. Analysis of the MammaPrint breast cancer assay in a predominantly postmenopausal cohort. Clin Cancer Res 2008;14:2988–93.

[106] Cardoso F, Piccart-Gebhart M, Van't Veer L, Rutgers E, TRANSBIG Consortium, The MINDACT trial: the first prospective clinical validation of a genomic tool. Mol Oncol 2007;1:246–51.

[107] Barker AD, Sigman CC, Kelloff GJ, Hylton NM, Berry DA, Esserman LJ. I-SPY 2: an adaptive breast cancer trial design in the setting of neoadjuvant chemotherapy. Clin Pharmacol Ther 2009;86(1):97–100.

[108] Pao W, Miller VA, Politi KA, Riely GJ, Somwar R, Zakowski MF, et al. Acquired resistance of lung adenocarcinomas to gefitinib or erlotinib is associated with a second mutation in the EGFR kinase domain. PLoS Med 2005;2:e73.

[109] Engelman JA, Jänne PA. Mechanisms of acquired resistance to epidermal growth factor receptor tyrosine kinase inhibitors in non-small cell lung cancer. Clin Cancer Res 2008;14:2895–9.

[110] Kosaka T, Yatabe Y, Endoh H, Yoshida K, Hida T, Tsuboi M, et al. Analysis of epidermal growth factor receptor gene mutation in patients with non-small cell lung cancer and acquired resistance to gefitinib. Clin Cancer Res 2006;12:5764–9.

[111] Kuang Y, Rogers A, Yeap BY, Wang L, Makrigiorgos M, Vetrand K, et al. Noninvasive detection of EGFR T790M in gefitinib or erlotinib resistant non-small cell lung cancer. Clin Cancer Res 2009;15:2630–6.

[112] Weisberg E, Manley PW, Cowan-Jacob SW, Hochhaus A, Griffin JD. Second generation inhibitors of BCR-ABL for the treatment of imatinib-resistant chronic myeloid leukaemia. Nat Rev Cancer 2007;7:345–56.

[113] Breccia M. Hematology: nilotinib and dasatinib – new 'magic bullets' for CML? Nat Rev Clin Oncol 2010;7:557–8.

[114] Fuerst ML. FDA approves dasatinib for imatinib resistance and intolerance 3 weeks after enthusiastic recommendation from ODAC. Oncol Times 2006;28:9–10.

[115] Kantarjian H, Shah NP, Hochhaus A, Cortes J, Shah S, Ayala M, et al. Dasatinib versus imatinib in newly diagnosed chronic-phase chronic myeloid leukemia. N Engl J Med 2010;362:2260–70.

[116] Breccia M, Alimena G. Nilotinib: a second-generation tyrosine kinase inhibitor for chronic myeloid leukemia. Leuk Res 2010;34:129–34.

[117] Cristofanilli M, Budd GT, Ellis MJ, Stopeck A, Matera J, Miller MC, et al. Circulating tumor cells, disease progression, and survival in metastatic breast cancer. N Engl J Med 2004;351:781–91.

[118] Somlo G, Lau SK, Frankel P, Hsieh HB, Liu X, Yang L, et al. Multiple biomarker expression on circulating tumor cells in comparison to tumor tissues from primary and metastatic sites in patients with locally advanced/inflammatory, and stage IV breast cancer, using a novel detection technology. Breast Cancer Res Treat 2011;128:155–63.

[119] Nagrath S, Sequist LV, Maheswaran S, Bell DW, Irimia D, Ulkus L, et al. Isolation of rare circulating tumor cells in cancer patients by microchip technology. Nature 2007;450:1235–9.

[120] Philip R, Carrington L, Chan. M. US FDA perspective on challenges in co-developing *in vitro* companion diagnostics and targeted cancer therapeutics. Bioanalysis 2011;3(4):383–9.

[121] Draft Guidance for Industry and Food and Drug Administration Staff – *In vitro* Companion Diagnostic Devices. Available from: <http://www.fda.gov/medicaldevices/deviceregulationandguidance/guidancedocuments/ucm262292htm>, accessed March 2013.

[122] Guideline for the manufacture of *in vitro* diagnostic products – food and drug administration center for devices and radiological health office of compliance. Available from: <http://www.fda.gov/downloads/medicaldevices/deviceregulationandguidance/guidancedocuments/ucm079101.pdf>, accessed March 2013.

[123] Moschos SA. Genomic biomarkers for patient selection and stratification: the cancer paradigm. Bioanalysis 2012;4(20):2499–511.

[124] Zieba A, Grannas K, Soderberg O, Gullberg M, Nilsson M, Landegren U. Molecular tools for companion diagnostics. N Biotechnol 2012;29:6.

[125] Hodgson DR, Whittaker RD, Herath A, Amakye D, Clack G, Biomarkers in oncology drug development. Mol Oncol 2009;3:24–32.

[126] Kim ES, Herbst RS, Wistuba II, Lee JJ, Blumenschein Jr GR, Tsao A, et al. The BATTLE Trial: personalizing therapy for lung cancer. Cancer Discov 2011;1:1.

[127] Rubin EH, Anderson KM, Gause CK. The BATTLE Trial: a bold step toward improving the efficiency of biomarker-based drug development. Cancer Discov 2011;1:17–20.

[128] Sequist LV, Muzikansky A, Engelman JA. A new BATTLE in the evolving war on cancer. Cancer Discov 2011;1:14–16.

[129] Kris M. Identification of drive mutations in tumor specimens from 1,000 patients with lung adenocarcinoma: The NCI's lung cancer mutation consortium. 2011 ASCO annual meeting. J Clin Oncol 2011:29 (June 20 Suppl.).

[130] US FDA. Table of pharmacogenomic biomarkers in drug labels. <www.fda.gov/Drugs/ScienceResearch/ResearchAreas/Pharmacogenetics/ucm083378.htm>, accessed March 2013.

[131] Chmielecki J, Peifer M, Jia P, Socci ND, Hutchinson K, Viale A, et al. Targeted next-generation sequencing of DNA regions proximal to a conserved GXGXXG signaling motif enables systematic discovery of tyrosine kinase fusions in cancer. Nucleic Acids Res 2010;38(20):6985–96.

3

Personalized Healthcare in Autoimmune Diseases

Cornelis L. Verweij[1], Brandon W. Higgs[2], Yihong Yao[2]

[1]Department of Pathology, VU University Medical Center, Amsterdam, The Netherlands
[2]MedImmune, LLC, Gaithersburg, Maryland

3.1 INTRODUCTION

One of the major challenges in the treatment of autoimmune diseases is the introduction of tailored therapies, or personalized medicine, to optimize the treatment efficacy for the individual patient, and to minimize the risk of adverse effects and unnecessary healthcare expenses. The concept of a personalized form of medicine has attracted interest in the search for clinical and molecular biomarkers to dissect responders from non-responders and to monitor drug efficacy. Mechanistically, there are two phases to explain the failure to respond to therapy with a biologic. Primary non-responders are characterized by a direct failure to reach the clinical endpoint, whereas the secondary non-responsiveness develops following a period of good response. One of the main reasons for the second phase of non-responsiveness can be attributed to the development of neutralizing antibodies, which takes at least six months to develop. It is believed that pathophysiological features at baseline and/or varying pharmacological effects among patients could be the underlying differences between a good and poor response in the primary phase.

Research efforts during the last decade have made progress in the identification of potential clinically relevant biomarkers to predict the primary non-response to biologic treatment and in monitoring drug efficacy. In this chapter we provide an overview of relevant biomarkers (predictive and pharmacodynamic markers) in the field of psoriasis, systemic lupus erythematosus (SLE), myositis, rheumatoid arthritis (RA), and multiple sclerosis (MS) which have potential clinical utility and provide a deeper understanding of the different molecular mechanisms underlying autoimmune diseases. These biomarkers have the potential to

*Y. Yao, B. Jallal, K. Ranade (Eds): Genomic Biomarkers for
Pharmaceutical Development.*
DOI: http://dx.doi.org/10.1016/B978-0-12-397336-8.00003-3

51

provide the right dose and right medications to patients with different molecular phenotypes of diseases.

3.2 INTERFERON GAMMA (IFN-γ) AS A PHARMACODYNAMIC (PD) MARKER FOR USTEKINUMAB (CNTO 1275) IN PSORIASIS

Skin-infiltrating lymphocytes expressing T helper type 1 cytokines, such as interferon gamma (IFN-γ), have been linked to the pathophysiology of psoriasis. Cytokines that elicit these immune responses include interleukin12 (IL12) and interleukin23 (IL23) produced by antigen presenting cells, both of which may represent potential therapeutic targets [1]. IL12, which has a substantial role in Th1 cell development from naïve T cells and IL23, known to activate Th17 cells, share a p40 subunit, which is activated in lesional skin of psoriatic plaques [2,3]. Both IL12 and IL23 were thought to be potentially involved in the pathogenesis of the disease in preclinical studies [4,5].

Briakinumab (ABT-874, Abbott) is a human IgG1 monoclonal antibody directed against the p40 subunit of IL12/IL23. Treatment with this molecule demonstrated improvement in Psoriasis Area and Severity Index (PASI-75) scores for all five treatment arms at week 12 in a Phase II dose-finding study [6]. Four Phase III studies evaluating briakinumab have shown superior response rates compared to patients treated with placebo, etanercept, or methotrexate, although one study showed an increase in major adverse cardiac events and malignancies in briakinumab-treated patients [7–11]. To date, briakinumab has not been approved by the Food and Drug Administration (FDA) for treatment of patients with psoriasis.

Ustekinumab (CNTO 1275, Centocor) is another human monoclonal antibody (IgG1κ) directed against the IL12/IL23 p40 subunit. In two Phase III trials (PHOENIX I and II) for moderate to severe psoriasis, at >76 weeks, ustekinumab was found to be both safe and effective [12,13]. A third Phase III trial (ACCEPT), compared the efficacy and safety of ustekinumab with etanercept in the treatment of moderate to severe plaque psoriasis [14]. This trial showed a significantly higher clinical response with ustekinumab over the 12-week study period compared to high-dose of etanercept [14]. Results also demonstrated the clinical benefit of ustekinumab among patients who failed to respond to etanercept [14]. Currently, ustekinumab is approved in Canada, Europe, and the United States for the treatment of moderate to severe plaque psoriasis.

In the case of development of ustekinumab, the investigators developed an effective PD marker to aid in the confirmation of mechanism of action (MoA). For example, in a Phase I study evaluating short-term safety, pharmacokinetics, pharmacodynamics, and clinical response of single subcutaneous (SC) administration of IL12/23 mAb in patients with moderate to severe plaque psoriasis, 20 patients were enrolled into dose cohorts of 0.27, 0.675, 1.35, and 2.7 mg/kg, or placebo and mRNA expression of type-1 cytokines or chemokines was measured from lesional skin biopsies both 24 hours before and 1 week following administration of IL12/23 mAb [15].

The transcripts included: TNF-α, IFN-γ, IL8, IL18, IL12/23 p40 subunit, IL23, p19 subunit, IL12 p35 subunit, IL10, interferon-inducible protein 10 (IP-10), regulated upon activation, normal T-cell expressed and secreted (RANTES), and CC chemokine ligand 2 (CCL2) [15]. It was observed that none of the mRNAs was significantly altered from baseline 1 week following treatment with either the IL12/23 mAb or placebo. However, when patients receiving IL12/23 mAb were stratified into groups of those who had a sustained PASI improvement (i.e., 70% average PASI improvement at weeks 8, 12, and 16; N = 4), and those who did not (N = 9), the sustained PASI improvers show significant decreases in mRNAs for the cytokines IFN-γ IL8, IL18 at week 1 post treatment, although the sample size was modest [15].

An additional non-randomized, open label, single dose, dose escalation Phase I trial confirmed the MoA of the intravenously administered human anti-IL12/IL23 p40 subunit mAb and demonstrated promising therapeutic efficacy in psoriasis [16,17]. Eighteen subjects with at least 3% body surface area involvement were enrolled in four dose groups that ranged from 0.1 to 5.0 mg/kg (0.1, 0.3, 1.0 and 5.0 mg/kg, respectively). Researchers used immunohistochemistry (IHC) and Real-Time Quantitative Reverse Transcription Polymerase Chain Reaction (qRT-PCR) to evaluate the histological and molecular changes of both cellular infiltrates and the same relevant type-1 cytokines and chemokines thought to play a role in the pathogenesis of psoriasis that were evaluated in the SC study, following a single administration of an anti-p40 mAb in psoriatic patients. At baseline, skin biopsies showed high infiltration of CD3+ T cells and CD11c+ dendritic cells into the lesion. At two weeks after administration of anti-IL12p40 mAb, the high responding patients (N = 9) showed a significant decrease from baseline (p = 0.025) in total CD3+ cells, unlike the low responding patients, indicating early signs of improvement at the cellular level in the high responders [16]. For mRNAs assayed in these patients, 17/18 patients treated with anti-IL12p40 mAb showed a significant reduction in IFN-γ at two weeks compared to baseline (p = 0.00001) [16]. Additional mRNAs that were significantly reduced at two weeks following administration of anti-IL12p40 mAb compared to baseline include IL8 (17/18 patients), IL10 (17/18 patients), IL12p40, IL23p19 (17/18 patients), TNF-α, IP-10, and MCP-1. When patients were stratified into high and low responders, the only major difference in mRNA patterns between baseline and two weeks post anti-IL12p40 mAb was with TNF-α, which failed to show a significant decrease in the low responder group [16]. The reduction of expression levels of these signaling molecules preceded clinical response and histological improvement [16]. IFN-γ is downstream of IL12 and the substantial decrease in its mRNA expression level two weeks following the administration of anti-IL12p40 mAb indicates that it could be used as a PD marker to evaluate inhibition of this molecule on IL12/IL 23-driven cellular activation, and could potentially be used for PK/PD modeling to determine the optimal dosing schedule to be used in the pivotal trial.

The baseline TNF-α mRNA expression showed a unique correlation to IFN-γ and RANTES and had a significantly higher expression level in high responders compared to low responders [16]. Additionally, a positive correlation was observed between baseline TNF-α mRNA level and clinical improvement at week 16 post drug treatment, as measured by the percent improvement of PASI [16]. This observation suggests the potential utility of baseline TNF-α mRNA level as a predictive marker to identify which plaque psoriatic patients might respond to anti-IL12p40 mAb therapy.

Taken together, the correlative studies in a Phase I trial confirmed that the anti-IL 12p40 mAb successfully inhibited the desired target, and provided positive, although early demonstration of therapeutic efficacy of the drug. This paved the way for the subsequent success of Phase II [18] and Phase III [7] clinical trials that followed that confirmed the therapeutic efficacy of the human anti-p40 subunit mAb in larger plaque psoriatic patient population.

3.3 A TYPE I INTERFERON SIGNATURE AS A PD MARKER AND POTENTIAL PREDICTIVE MARKER FOR ANTI-TYPE I INTERFERON THERAPY IN SYSTEMIC LUPUS ERYTHEMATOSUS (SLE), MYOSITIS AND SYSTEMIC SCLEROSIS (SSc)

The type I interferon (IFN) family consists of multiple members, including types α (with 14 subtypes), β, ε, κ, ω, δ, and τ. Biological functions range from viral or bacterial infection defense to immunomodulation and anti-proliferation activity. Because type I IFNs, specifically IFN-α/β, have such a range of functional influence on immune-mediated activity, they have been investigated for decades for their roles in indications such as inflammation, autoimmunity, and cancer [19–22].

The type I IFN receptor (IFNAR), a heterodimer of IFNAR-1 and IFNAR-2, is activated by the binding of type I IFN, resulting in the conformational changes of IFNAR and activation of the Janus kinase (JAK)/signal transducers and activators of transcription (STAT) signaling pathway [23–25]. This in turn initiates a cascade of alternative processes, eventually leading to transcription activators binding the IFN response element (ISRE) and the production of IFN-inducible genes. The activation of this type I IFN pathway has been quantified using the expression level of these IFN-inducible genes with high sensitivity in multiple rheumatic diseases [26–30].

Sifalimumab, an investigational human IgG1κ mAb that binds to most of the IFN-α subtype with high affinities, has been evaluated in four separate clinical trials studying adult patients with SLE or myositis (dermatomyositis [DM] or polymyositis [PM]) [13–17] (NCT00299819, NCT00482989, NCT00657189, and NCT00533091). All four trials were randomized, double-blinded, placebo-controlled and dose escalation studies, with three using intravenous dosing and one, subcutaneous dosing of the drug. A type I IFN-inducible gene signature was developed and used to evaluate the PD of this molecule in both SLE and myositis patients. This gene signature, consisting of 13 or 21 genes in myositis or SLE respectively, was shown to be neutralized to different degrees in blood by sifalimumab in SLE and both blood and muscle tissue in myositis patients in the aforementioned clinical trials [31–34]. The degree of neutralization of this signature in patients of varying disease activities might be due to the type of type I IFNs present in those patients. This hypothesis needs further confirmation from future trials. Figure 3.1 and Figure 3.2 illustrate the target neutralization of the type I IFN signature in the blood in a Phase Ia clinical trial at dose levels of 1 (N = 6), 3 (N = 6), 10 (N = 7), and 30 (N = 8) mg/kg (plus placebo; N = 17) in SLE and in both blood and muscle tissue in a Phase Ib clinical trial at dose levels of 0.3 (N = 7), 1 (N = 8), 3 (N = 16), and 10 (N = 8) mg/kg (plus placebo [N = 12]) in myositis patients, respectively.

Although this type I IFN signature has been used to evaluate the drug-target engagement in the above mentioned clinical trials, there exists a subgroup of either SLE or myositis patients that do not show substantial target neutralization by sifalimumab. Besides the possibility of dose, it could be that other type I IFNs, besides IFN-α, were the major contributor to elevated type I IFN signature observed in these patients. As noted previously, sifalimumab has specificity for the majority of the IFN-α subtypes, but not IFN-β, which can signal through the same IFNAR and activate the type I IFN pathway. The prevalence of IFN-β compared to IFN-α, and its contribution to disease pathogenesis in myositis, is not clear. The IFN-β level in the blood has been shown to both be unique to and correlate with an elevated IFN signature in DM patients [35]. One group has shown that IFN-β and not IFN-α transcripts are over-expressed

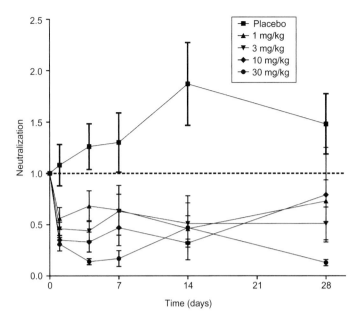

FIGURE 3.1   Effects of sifalimumab on type I IFN signature in subjects with SLE. Level of expression of 21 type I IFN-inducible genes for each subject relative to baseline expression. For this plot, SLE subjects must have had an overexpression of the gene signature of at least threefold, normalized to 1 at baseline. The change from baseline mean±SE level of signature gene expression in whole blood after sifalimumab treatment is shown. Maximum neutralization of the gene signature occurred within 24 hrs after treatment. Hotelling's T2 test with pairwise comparisons up to 14 days after dosing was used for this analysis. *Courtesy of Ann Rheum Dis [32].*

in polymyositis (PM) patients as well as dermatomyosistis (DM) and juvenile DM muscle specimens from patients, which suggests that there is a local source of IFN-β that might contribute to the lack of stronger target neutralization by sifalimumab in muscle compared to blood [36]. Type III IFN (IFN-λ) can also bypass the PD effect of sifalimumab and induce activation of IRF-9 and transcription of IFN-inducible genes [37], although IFN-λ induces a much smaller effect on type I IFN-inducible genes in blood. One interesting question that stemmed from these observations is whether the neutralization of the type I IFN signature in SLE and myositis ultimately showed positive correlation with clinical response. MEDI-546, an investigational human IgG1κ mAb directed against subunit 1 of the IFNAR, has been evaluated in an open label single- and multiple-dose Phase Ia clinical trial in patients with systemic sclerosis (SSc) (NCT00930683) and is currently under investigation in a Phase IIb trial in patients with SLE (NCT01438489). A five gene, type I IFN signature was used to assess the PD of MEDI-546 in SSc patients, similar to that conducted with sifalimumab in both SLE and myositis patients [38]. The results from the trial (N = 34) indicated that the type I IFN signature was neutralized in a dose-dependent manner in SSc patients following both single- and multiple-dose regimens of MEDI-546 [38]. In the high-dose 20.0 mg/kg single-dose cohort and the 1.0 mg/kg and 5.0 mg/kg multiple-dose cohorts, median-fold change scores generally remained fully neutralized through day 84 (Fig. 3.3).

The potential clinical utility of the type I IFN signature as a potential predictive marker for anti-type I IFN therapy, especially for those that target the IFNAR where complete target

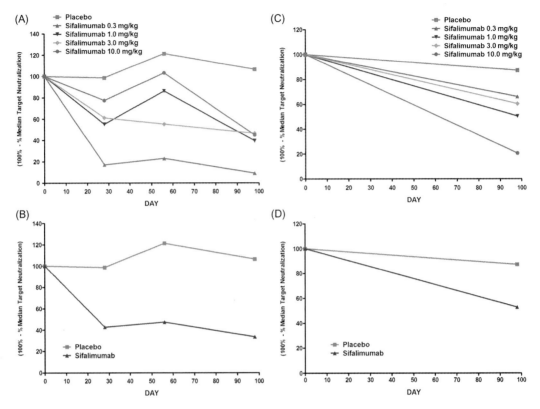

FIGURE 3.2   Median target neutralization of the type I IFN signature as calculated based on the expression of 13 genes pre dose and post dose up to day 98 for dose cohorts of 0.3 mg/kg (N = 7), 1.0 mg/kg (blue; N = 8), 3.0 mg/kg (N = 16), 10 mg/kg (N = 8), and placebo (N = 12) in (A) blood and (C) muscle specimens, as well as median combined dose cohorts (blue) versus placebo treatment cohorts (red) in (B) blood and (D) muscle specimens from dermatomyosistis (DM) or polymyositis (PM) patients. The y axis represents the percentage of IFNGS remaining following treatment; each line is the median of the respective dose cohort. *Courtesy of Ann Rheum Dis [34].*

neutraliation could be achieved, needs to be carefully evaluated in larger Phase IIb trials. Such trials will also provide more evidence of whether type I IFN, and what subtypes of type I IFN, might be involved in the pathogenesis of rheumatic diseases such as SLE and myositis.

3.4 BIOMARKER FOR ANTI-CD20 mAb THERAPY IN PATIENTS WITH RHEUMATOID ARTHRITIS (RA)

3.4.1 Rituximab Depletes Circulating B Cells and Shows Clinical Efficacy in Patients with RA

B cells manifest from the bone marrow and once activated can differentiate into either plasma or memory forms – either through direct processes or additional intermediate stages [39]. The B

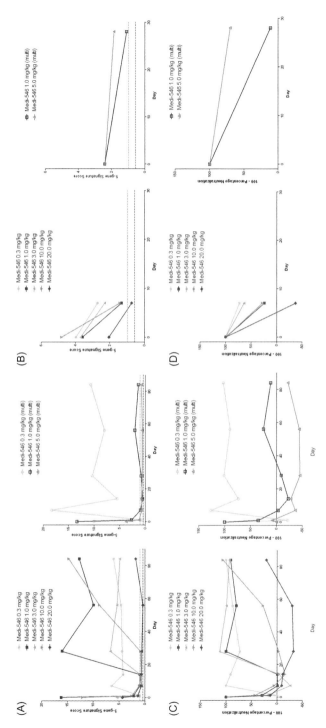

FIGURE 3.3 Median type I IFN (A and B) GS profiles and (C and D) percent remaining GS in diffuse SSc patients following single or multiple IV administrations of MEDI-546 in (A and C) whole blood specimens or (B and D) skin specimens from MI-CP180 trial. For each pair of plots, the single and multiple-dose treatment cohorts have been separated into their respective graph. The x axis represents time from the start of the study in days, where day 0 is pre dosing. Target modulation for each dose cohort is reported as a percentage from starting values of 100%, so each point post treatment for each cohort indicates the median percentage of remaining gene signature (GS). Only baseline positive GS_0 score SSc patients were plotted. The minimum and average GS score among the pool of normal healthy controls are shown as the black dashed line and red dashed line respectively (A and B). *Courtesy of Clin Pharm Ther [38].*

cell antigen CD20 is a 35–37 kD tetra-spanning integral membrane phosphoprotein that is highly expressed by naïve, mature, and memory B cells, but not by precursor B cells and antibody-producing plasma cells. This surface antigen is not shed by the cell and there are no detectable levels of a soluble form in the serum [40]. Since this antigen is also expressed on the surface of neoplastic B cells, it is an attractive target for B cell lymphoma and leukemia [40]. In addition to these indications, RA, a chronic, disabling autoimmune disease characterized by synovial joint arthritis and fatigue, has B cell involvement in its pathogenesis, associated with autoantibody production leading to autoantibody/autoantigen complex deposits and tissue injury.

Rituximab, developed by Biogen Idec, a chimeric mouse/human IgG1 mAb, is a CD20 antagonist that depletes CD20 naïve and memory B cells from the blood, bone marrow, and lymph nodes by a mixture of apoptosis, antibody-dependent cell-mediated cellular cytotoxicity (ADCC), and complement-dependent cytotoxicity (CDC) [41]. This molecule has demonstrated efficacy in patients with RA who do not respond to anti-TNF mAb therapies, and has been approved by the FDA for this patient population, in addition to patients with diffuse large B cell lymphoma and chronic lymphocytic leukemia [42]. CD20+ B cell depletion (mean decrease of 97%) in RA is essentially complete at one month after the start of a single treatment and sustains for several months [43–46]. Peripheral B cells repopulate to almost baseline levels between 6 and 10 months after treatment [47,48]. Repopulation starts with the appearance of CD5+/CD38^{high} naïve B cells, followed by an increase in immature CD19-/IgD-/CD38^{high}/CD10^{low}/CD24^{high} B cells [46]). Serum titers of rheumatoid factor (RF) and anti-citrullinated protein antibodies (ACPA) significantly decreased at 24 and 36 weeks, respectively, whereas total immunoglobulin levels and antibody titers against recall antigens were not affected. This finding may suggest that rituximab selectively affects short-lived autoantibody-secreting plasma cells [47–49]. The long-lasting depletion of peripheral blood B cells initially raised concerns with respect to long-term safety. As it stands, the safety of rituximab is comparable to that of other biologic Disease-modifying antirheumatic drugs (DMARDs). Adverse events were mild to moderate infusion reactions observed in 2% of the rituximab-treated patients. Ocrelizumab is a humanized mAb that is the second generation antagonist for CD20 from Genentech and Biogen, and this is currently being evaluated in multiple sclerosis in late stage clinical trials [42].

Despite the effective depletion of circulating B cells in nearly all treated patients, a substantial percentage of patients do not respond to rituximab treatment. In order to effectively utilize rituximab and prevent unnecessary costs, risk of adverse effects, and delays in applying effective treatment, it is necessary to restrict treatment to only those patients who will benefit.

3.4.2 IgJ mRNA as a Predictive Marker of Non-response to Rituximab Therapy in Patients with RA

A single transcript for IgJ has recently been developed as a biomarker to predict which RA patients will show reduced clinical response to rituximab therapy [42]. In a *post hoc* analysis, this biomarker was shown to be a consistent baseline predictor of poor response in RA patients dosed with either rituximab or ocrelizumab, unlike those in patients administered with placebo from four different clinical trials – DANCER, SERENE (NCT00299130), REFLEX (NCT00468546), and SCRIPT (NCT00476996). RT-qPCR and flow cytometry assays were performed on whole blood specimens from RA patients dosed with rituximab, ocrelizumab, or placebo (depending on the clinical trial) at baseline and post dose time points. The specific mRNAs interrogated included CD19, CD20, FCRL5, BCMA, and IgJ, with IgJ showing the best single gene performance [42].

Using the REFLEX clinical trial as a training set and the American College of Rheumatology 50% improvements criteria (ACR50) as the clinical response measurement, the transcript for IgJ was used to partition RA patients into a high or low group using 0.1 as a cut point. This subgroup analysis showed a 22% difference in ACR50 response rates between the high and low drug-administered patient groups – a result that was not reproduced in the placebo cohort. Additionally, this difference between high and low IgJ mRNA patient groups dosed with anti-CD20 mAb was reproduced in three additional clinical trials (Fig. 3.4) [42]. The subgroup identified by this biomarker had a patient prevalence of 25%, indicating the approximate population size that may benefit from this baseline predictive biomarker.

3.4.3 RA Patients with High Baseline Type I Interferon Signature are Less Likely to Respond to Anti-CD20 mAb Therapy

Genome-wide gene expression profiling has identified that the type I IFN signature constitutes another biomarker for predicting non-responders to anti-CD20 mAb therapy in RA patients [50]. Good responders have a low or absent IFN response activity at baseline, whereas non-responders display an activated type I IFN-system before the start of treatment. The association between baseline type I IFN levels and clinical response is in line with previous findings, wherein it was demonstrated in two different cohorts that patients with a low IFN signature had a significantly greater reduction in the DAS28 and more often achieved a European League Against Rheumatism (EULAR) response at weeks 12 and 24 [51]. Evidence for the clinical utility of the IFN signature genes as biomarker to predict the non-response outcome of rituximab treatment in RA was demonstrated using Receiver Operator Characteristics (ROC) curve analysis. ROC-curve analysis using an optimal IFN type I gene set (3–5 genes) as predictor in an independent group of 26 RA patients treated with rituximab revealed very good clinical utility, reflected by an area under the curve of 0.87 (p = 0.001) according to ΔDAS28 response criteria (Fig. 3.5) [50].

Results from these studies suggest that IFN^{high} RA patients represent a different pathogenic subset of RA marked by a failure to respond to B cell depletion therapy. A simple explanation could be that the pathogenesis in IFN^{high} patients is less dependent on B cells, compared to IFN^{low} patients. Alternatively, a high baseline IFN activity may be associated with the presence of a subset of pathogenic B cells insensitive to the effects of rituximab. These could be present at baseline and could survive in synovial or bone marrow tissues due to, for example, incomplete B cell depletion effectors or concomitant expression of B cell survival factors such as BAFF/BLyS [52]. IFNs may also affect B cell differentiation, such as *in situ* differentiation in CD20- plasma blasts [52].

These findings provide a basis for the clinical utility of the IFN signature as a biomarker for prediction of clinical response to rituximab, and constitute a step towards patient-tailored treatment in RA. This biomarker could also apply to other indications, such as multiple sclerosis and SLE, which benefit from B cell depletion. Further research is necessary to validate the clinical utility of this biomarker, eventually in combination with other biomarkers and/or clinical variables, in a multicenter setting using prospective studies. Ultimately, these results may provide the basis for a method that can be implemented in clinical practice to meet the unmet need to prescribe the most effective therapy for a particular patient.

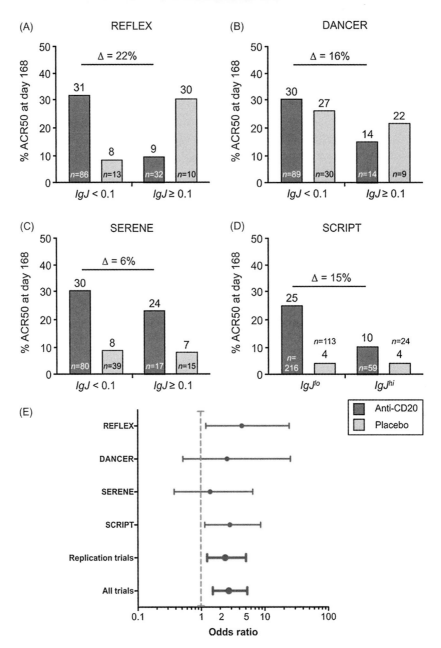

FIGURE 3.4 An RA subgroup defined by an IgJ mRNA biomarker with reduced efficacy after anti-CD20 therapy. (A) Baseline mRNA samples from the REFLEX trial of rituximab in RA were used to identify optimal biomarker thresholds for IgJ as a predictor of ACR50 response rates at 6 months (day 168) for subjects treated with anti-CD20 (blue bars) or placebo (yellow bars). (B and C) This biomarker threshold (IgJ $\geq$0.1 or <0.1 expression unit) was then tested prospectively in baseline mRNA samples from the DANCER (B) and SERENE (C) trials of rituximab in RA. (D) Biomarker thresholds for the SCRIPT trial of ocrelizumab in RA were based on percentage thresholds from the rituximab studies (see text for details). Δ in (A) to (D) denotes the ACR50 percentage difference for the active anti-CD20 arm between the IgJ^{hi} and the IgJ^{lo} subgroups. ^{n} refers to the number of individual subjects in each subgroup, and the number above the bars is the percent ACR50 for each subgroup. (F) ORs and 95% CI for the enrichment of ACR50 responses in the IgJ^{lo} subgroup compared to the IgJ^{hi} subgroup for the individual trials, the replication trials in aggregate (DANCER, SERENE, and SCRIPT), and all trials together are depicted.

GENOMIC BIOMARKERS FOR PHARMACEUTICAL DEVELOPMENT

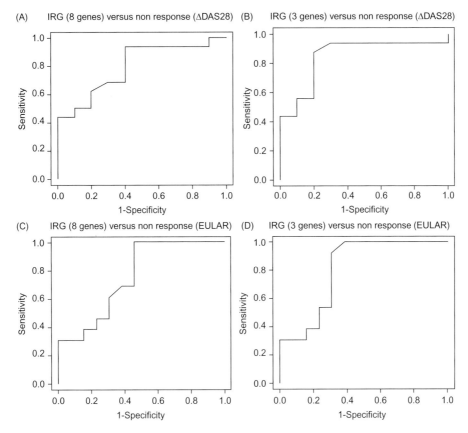

FIGURE 3.5 Receiver operating characteristics (ROC) curves for the IFN response genes as predictor for non-response upon rituximab treatment in the RA validation cohort (N = 26). (A) AUC (0.82) for the eight IFN response gene set based on ΔDAS28 response criteria, (B) AUC (0.87) for the three IFN response gene set based on ΔDAS28 response criteria, (C) AUC (0.78) for the eight IFN response gene set based on EULAR response criteria (responders and intermediate responders vs. non-responders) and (D) AUC (0.83) for the three IFN response gene set based on EULAR response criteria (responders and intermediate responders versus non-responders). On the y axis sensitivity and on the x axis 1-specificity is indicated. AUC, area under the curve; DAS28, 28 joints disease activity score; EULAR, European League against Rheumatism; IRG, interferon response gene; RTX, rituximab. *(Adapted with permission from Raterman et al. [50].)*

3.4.4 A Type I Interferon Signature as a Pharmacodynamic Marker of Anti-CD20 mAb Therapy in Patients with RA

Vosslamber and colleagues studied the pharmacological effects of rituximab in RA and observed a difference in the expression of IFN type I genes during rituximab treatment that distinguishes responders from non-responders (Fig. 3.6) [53,54]. Responders exhibited an increase in IFN response activity after three months' treatment with rituximab, whereas the IFN response activity remained stable during treatment in the non-responders. This means that good responders have a low or absent IFN response activity at baseline and develop

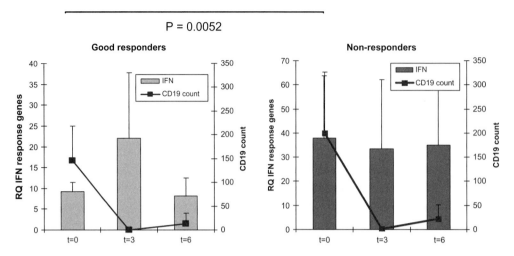

FIGURE 3.6   Pharmacodynamics of the interferon (IFN) type I-response activity during rituximab treatment reveal marked differences between responders and non-responders. Shown are pharmacodynamic measurements in 13 patients with rheumatoid arthritis (RA) of a set of six type I IFN response genes (IFI44, IFI44L, HERC5, RSAD2, LY6E, and Mx1) (left y axis) and B cell counts based on CD19 cytometry (right y axis) at baseline (t0), 3 (t3) and 6 months (t6). Patients were stratified in responders and non-responders based on changes in disease activity score (ΔDAS) criteria. Pharmacodynamic analyses of the IFN type I-response activity during rituximab treatment revealed marked differences between responders and non-responders for baseline IFN response activity (ΔDAS p = 0.0052) and the rituximab induced increase in IFN type I activity at 3 months (ratio t3/t0) (ΔDAS p = 0.049). The change in IFN response gene activity during rituximab treatment negatively correlated with the corresponding baseline level, although no significance was reached (p = 0.0576; R = −0.53). (Adapted with permission from: Vosslamber S. et al. [53].)

IFN response activity to a level comparable with that of non-responders during three months of treatment, whereas non-responders displayed an activated type I IFN-system already before the start of treatment, which remains stable during treatment. The IFN signature score returned to baseline values at six months after the start of treatment (Fig. 3.6).

Thus a pharmacological increase in IFN response activity during rituximab treatment is associated with a favorable response and may provide insight into the biological mechanism underlying the therapeutic response. The dynamic increase in type I IFN activity might be a determining factor in the ameliorative effect of B cell depletion therapy in RA, and might explain the increased BAFF/BLyS levels, persistence of pathogenic B cells, and the change in macrophage function. These findings provide a basis for further study on the role of the IFN signature as a biomarker for effective dosing and timing of treatment, towards patient-tailored treatment in RA.

A dynamic increase in IFN response activity with concomitant B cell depletion may be a prerequisite for a beneficial outcome. This hypothesis may also explain the beneficial effects of rituximab treatment observed in multiple sclerosis, a disease that responds favorably to effects of IFN-β. Conversely, a pharmacological increase in the type I IFN activity by rituximab may lead to disease progression and/or an increase in disease activity in type I IFN driven diseases such as SLE, and may explain the failure to meet clinical endpoints in recent randomized, placebo-controlled trials of rituximab [55]. The data suggest that rituximab might be less effective in those SLE patients who experience an increase in their type I IFN response activity levels during rituximab treatment.

3.5 A TYPE I INTERFERON SIGNATURE AS A PREDICTIVE MARKER OF RESPONSE TO INTERFERON-β THERAPY IN PATIENTS WITH MULTIPLE SCLEROSIS

Multiple sclerosis (MS) is the most common inflammatory disease of the central nervous system (CNS), and is characterized by progressive neurological dysfunction due to demyelination of the nerves [56]. Destruction of the myelin sheet disrupts the nerve action potential, leading to muscle weakness, blurred vision and motor symptoms. Clinical experience has revealed that MS is a heterogeneous disease that is difficult to treat. The heterogeneous nature within the different forms of MS is reflected by the clinical presentation, which ranges from mild to severe demyelinating disease. The existence of heterogeneity in the brain of MS patients is proposed to represent different pathogenic processes underlying MS, ranging from an inflammation-mediated to an immune-independent demyelinating form of brain tissue destruction. The heterogeneity is also reflected by the considerable variation in responsiveness to treatment and most likely has its origin in the multifactorial nature of the disease, whereby specific combinations of genetic risk factors together with an appropriate environmental trigger influence not only susceptibility but also the severity, pathogenesis and outcome of the disease.

No curative therapy is currently available, and the majority of affected individuals are ultimately disabled. Therapies are mainly directed to protect CNS cells, modulate T cells and induce remyelination. IFN-β was the first agent to show clinical efficacy in the most common form of MS, i.e., relapsing remitting (RR) MS. According to results from clinical trials, IFN-β treatment reduces relapse rates by about 30%, decreases the formation of inflammatory lesions in the CNS, extends remission periods and possibly slows down progression of disability [57]. Currently, a prolonged course of IFN-β treatment is still the best available therapy for RRMS. There are several IFN-β preparations available for therapy (i.e., recombinant IFN-β1b produced in *Escherichia coli* (Betaseron®/Betaferon® by Bayer and Extavia® by Novartis), and recombinant IFN-β1a derived from Chinese hamster ovary (CHO) cells (Avonex® by Biogen Inc., Rebif® by Merck Serono S.A., and Pfizer and Cinnovex® by Cinnagen).

Unfortunately, clinical experience indicates that there are IFN 'responders' as well as 'non-responders'. A high proportion of about 40% of the patients do not or only poorly respond to IFN-β treatment [58]. Moreover, IFN-β therapy is associated with adverse reactions such as flu-like symptoms and transient laboratory abnormalities. Given the destructive nature of RRMS, the risk of adverse effects, and considerable costs for therapy, there is a strong need to make predictions of success before the start of therapy.

Mechanistically, there are two phases to explain the failure to respond to therapy with a biological agent. The first phase of unresponsiveness is the consequence of a direct failure of the agent to suppress the disease, whereas the second phase may follow a period of good response. The first phase implies that baseline characteristics and/or pharmacodynamic responses may differ between patients, leading to interindividual differences in clinical efficacy, which relate the heterogeneous nature of RRMS. The second phase can be explained by immunogenicity, i.e., the development of neutralizing antibody (NAb) directed against the drug. However, not all non-responders develop NAbs, and if induced it takes at least six months for NAbs to develop, which can also disappear again over time [59,60].

Efforts to understand differential responsiveness have focused primarily on the mechanistic (that is, the primary) phase of unresponsiveness. Van Baarsen and colleagues were the first to report that the baseline IFN type I signature determined the pharmacological response to IFNβ [61,62]. The results revealed that an increased IFN response gene activity at baseline was associated with the absence of a pharmacological effect of IFNβ-treatment in RRMS. Comparing baseline expression levels of an optimal performing set of 15 type I IFN response genes with pharmacological response outcome revealed a significant negative correlation (R = −0.7208; p = 0.0016). The negative correlation between pharmacological response and baseline levels of IFN-induced genes is consistently observed over time, at one, three and six months after the initiation of the therapy and was confirmed in an independent cohort of 30 RRMS patients (p < 0.0085 and R = −0.4719) (Fig. 3.7). These results were confirmed by other investigators, who observed that an increase in the expression of type I IFN genes was much lower or absent in the IFN[high] RRMS patients [63–65].

Given the ameliorative role of IFN-β on disease activity in RRMS, these results appear counterintuitive. One possible explanation for these observations is that the type I IFN activity in the IFN[high] RRMS patients was already saturated prior to the initiation of the therapy, resulting in the absence of a pharmacological effect of administered IFN-β in these patients. Indeed, it was demonstrated that peripheral blood mononuclear cells (PBMC) of IFN[high] RRMS patients had lost the capacity to respond to *in vitro* stimulation with IFN-β, consistent with the *in vivo* findings [62]. Moreover, Zula and colleagues reported that IFN-α/β therapy in patients with RRMS induced the expression of a STAT dependent gene signature in myeloid cells of responders only [66]. Others observed that the IFN[high] status correlated with endogenous IFN-β and elevated IFN receptor 1 (IFNAR1) expression by monocytes [63]. Also the capacity of IFN-β to induce its own expression was deficient in cells from IFN[high] patients [64].

Overall the data convincingly support the conclusion that the *in vivo* pharmacological response to IFN-β treatment is dependent on the intrinsic type I IFN response gene activity status prior to the start of the therapy [67]. Obviously this could also have consequences for the clinical outcome of IFN-β therapy. Indeed, baseline type I IFN response gene expression levels were predictive of the clinical outcome of IFN-β treatment in RRMS [63,68]. Responders were defined as having no increase in the Expanded Disability Status Scale (EDSS) score and no relapses during the 24 month follow-up period. Non-responders had at least one relapse and one point increase in EDSS score during that period. As anticipated from the pharmacological outcome study, Comabella and colleagues observed that non-responders were characterized by an increased expression of type I IFN response genes at baseline. The predictive accuracy of a gene set, consisting of predominantly type I IFN genes, reached 78%. These findings were replicated in an independent group of RRMS patients (predictive accuracy 63%). The genes that were selectively induced by type I IFN were found to be the best predictors of efficacy. These results warrant further studies to validate the clinical utility of the IFN signature as a biomarker to predict the response to IFN-β treatment in RRMS [63].

Accordingly, baseline phospho-STAT1 levels were found to be significantly higher in monocytes derived from non-responders when compared to responders. Upon stimulation of the monocytes with IFN-β, no differences between the phospho-STAT1 levels were observed between responders and non-responders [63]. These findings are consistent with previous

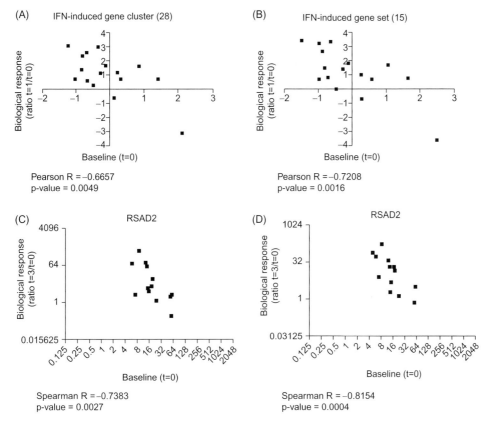

FIGURE 3.7 Correlation between baseline and pharmacological response to IFN-β therapy. Pharmacological responses were calculated for the indicated follow-up periods relative to the baseline expression levels using a set of 28 (A) or 15 (B) IFN-induced genes or RSAD2 as a single IFN-induced gene (C and D). The results reveal in significant negative correlations between the pharmacological response at 1, 3 and 6 months after the start with IFN-β treatment for the measurement of IFN response gene activity (IFN signature expression). In C and D the expression levels of RSAD2 are measured by quantitative real-time PCR and normalized to the expression levels of GAPDH. (A) Pharmacological response after 1 month using a 28 gene IFN cluster; (B) Pharmacological response using a selection of 15 IFN response genes; (C) Pharmacological response after three months, using RSAD2 gene expression levels; (D) Pharmacological response after six months using RSAD2 gene expression levels. *(Adapted with permission from: van Baarsen LG, et al. [62].)*

results, which revealed that in the IFN^high patients the type I IFN pathway is fully activated prior to therapy and cannot be activated further [62]. Remarkably, LPS (via TLR4) triggered induction of IFN-α was significantly increased in non-responders compared to responders. This effect is most likely also mediated by the monocytes and suggests an uncoupling of IFN-α and IFN-β production.

In accordance with an increased IFN activity at baseline, Axtell and colleagues reported that non-responders had a significantly elevated serum concentration of IFN-β compared to the responders [69]. High IFN-β concentration correlated with increased IL17F in the serum suggesting a tight biological association between these two cytokines. One hypothesis for the

TABLE 3.1   Association between IRF5 Genotypes and Clinical Response Outcome Measures

	Genotype vs Pharmacological Response		Genotype vs MRI-Based Lesion Load		Genotype vs MRI-Based Non-Responders (One or more new T2 Lesions)		Genotype vs Time to First Relapse	
	rs2004640 TT	rs4728142 AA	rs2004640 TT	rs4728142 AA	rs2004640 TT	rs4728142 AA	rs2004640 TT	rs4728142 AA
VUmc patients (n = 30)	$P = 0.0006$[a]	$P = 0.0023$[a]					NA	NA
VUmc + CEM-Cat patients (n = 28 + 45)	NA	NA	$P = 0.003$[b] $P = 0.013$[c]	$P = 0.103$[b] $P = 0.201$[c]	$P = 0.010$[b] $P = 0.073$[c]	$P = 0.154$[b] $P = 0.440$[c]		
BWH + VUmc validation cohort (24 months follow-up)	NA	NA	NA	NA	NA	NA	$P = 0.037$	NS
BWH validation cohort (long follow-up)	NA	NA	NA	NA	NA	NA	$P = 0.030$	$P = 0.067$

[a]*Student's t test.*
[b]*Kruskal–Wallis.*
[c]*Mann–Whitney.*
(Adapted with permission from: Vosslamber S; van der Voort LF; van den Elskamp IJ; Heijmans R; Aubin C; Uitdehaag BM, et al. Interferon regulatory factor 5 gene variants and pharmacological and clinical outcome of Interferon beta therapy in multiple sclerosis. Genes Immun. 2011; 12: 466–72.)
Abbreviations: BWH, Brigham and Women's Hospital in Boston; CEM-Cat, Centre d'Esclerosi Múltiple de Catalunya in Barcelona; NA, not applicable; NS, not significant; VUmc, VU University medical center. P-values < 0.05 are shown in bold.

observation was that non-responders have aggressive Th17 cells reflected by the increase in IL17F production, and IFN-β was produced to counteract inflammation. Alternatively, endogenous IFN-β acts in a pro-inflammatory manner during Th17 cell driven disease. However, the role of IL17F as predictor of the response to IFN-β could not be confirmed in an independent cohort [70].

Genetic studies revealed a contribution for variants in the gene encoding Interferon Regulatory Factor 5 (IRF5), a master regulator of the IFN/TLR pathway. IRF5 is a transcription factor that functions as a central mediator of Toll-like receptor signaling and is involved in the production of type I IFN, apoptosis, cell-cycle regulation, cell adhesion and pro-inflammatory reactions [71,72]. Moreover, expression of IRF5 is induced after activation of the IFN type I receptor, indicative that IRF5 is not only important in the production of type I IFN, but also in the regulation of IFN type I-induced gene activity [73]. Genetic variation in the IRF5 gene has been found to be strongly associated with SLE, a disease wherein type I IFNs are clearly associated with disease activity and severity, and IFN response gene activity [74–76].

Vosslamber and colleagues found that RRMS patients with the IRF5 rs2004640-TT and rs47281420-AA genotype showed a poor pharmacological response to IFN-β compared with patients carrying the respective G-alleles (p = 0.0006 and p = 0.0023, respectively) (Table 3.1,

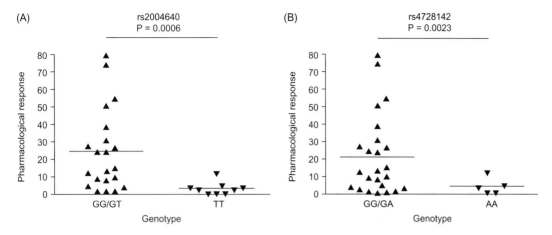

FIGURE 3.8 Relationship between *IRF5* genotypes and pharmacological response to IFN-β treatment. Pharmacological response at 1 month following the start of IFN-β treatment was determined in a cohort of 30 patients with RRMS and the association with rs2004640 (A) and rs4728142 (B) genotypes was determined. Pharmacological responses are lower in patients homozygous for the rs2004640 T allele (A) or rs472814 A-allele (B) compared to other genotypes. *(Adapted with permission from: Vosslamber S; van der Voort LF; van den Elskamp IJ; Heijmans R; Aubin C; Uitdehaag BM, et al. Interferon regulatory factor 5 gene variants and pharmacological and clinical outcome of Interferon beta therapy in multiple sclerosis. Genes Immun. 2011; 12: 466–72.)*

Fig. 3.8) [53]. Moreover, patients with the rs2004640-TT genotype developed more magnetic resonance imaging (MRI)-based T2 lesions during IFN-β treatment (p = 0.003). Accordingly, an association between MRI-based non-responder status and rs2004640-TT genotype was observed (p = 0.010). The clinical relevance of the rs2004640-TT genotype was validated in an independent cohort wherein a shorter time to first relapse was found (p = 0.037). These findings suggest a role for IRF5 gene variation in the pharmacological and clinical outcome of IFN-β therapy that might have relevance as a biomarker to predict the response to IFN-β in RRMS.

Altogether the above findings indicate that the presence of the type I IFN signature defines two clinically distinct subsets of MS patients, based on the clinical response outcome of IFN-β therapy. The IFN[high] patients represent a group of patients who express a failure to demonstrate a pharmacological response to IFN-β treatment and consequently do not show a clinical response. IFN[low] patients exert a pharmacological response with concomitant clinical response. The published results warrant further studies to validate the clinical utility of the IFN signature as biomarker to predict the response to IFN-β treatment. Due to the temporal aspects related to monitoring of the clinical response, it remains to be determined whether research findings from studies on the primary phase of unresponsiveness might be intimately linked to processes that are (also) related to NAb development.

Acknowledgments

Supported in part by the Center for Medical Systems Biology (a center of excellence approved by the Netherlands Genomics Initiative/Netherlands Organization for Scientific Research), the Center for Translational and Molecular Medicine (CTMM) (an initiative from the ministry of Economic Affairs of The Netherlands) and grants from the Dutch Arthritis Foundation and MS Research (04–549 MS and 08–660 MS).

References

[1] Trinchieri G. Interleukin-12 and the regulation of innate resistance and adaptive immunity. Nat Rev Immunol 2003;3(2):133–46.

[2] Morrison PJ, Ballantyne SJ, Kullberg MC. Interleukin-23 and T helper 17-type responses in intestinal inflammation: from cytokines to T-cell plasticity. Immunology 2011;133(4):397–408.

[3] Kleinschek MA, Muller U, Schutze N, et al. Administration of IL-23 engages innate and adaptive immune mechanisms during fungal infection. Int Immunol 2010;22(2):81–90.

[4] Hong K, Chu A, Ludviksson BR, Berg EL, Ehrhardt RO. IL-12, independently of IFN-gamma, plays a crucial role in the pathogenesis of a murine psoriasis-like skin disorder. J Immunol 1999;162(12):7480–91.

[5] Lee E, Trepicchio WL, Oestreicher JL, et al. Increased expression of interleukin 23 p19 and p40 in lesional skin of patients with psoriasis vulgaris. J Exp Med 2004;199(1):125–30.

[6] Kimball AB, Gordon KB, Langley RG, Menter A, Chartash EK, Valdes J. Safety and efficacy of ABT-874, a fully human interleukin 12/23 monoclonal antibody, in the treatment of moderate to severe chronic plaque psoriasis: results of a randomized, placebo-controlled, Phase II trial. Arch Dermatol 2008;144(2):200–7.

[7] Gottlieb AB, Leonardi C, Kerdel F, Mehlis S, Olds M, Williams DA. Efficacy and safety of briakinumab vs. etanercept and placebo in patients with moderate to severe chronic plaque psoriasis. Br J Dermatol 2011;165(3):652–60.

[8] Strober BE, Crowley JJ, Yamauchi PS, Olds M, Williams DA. Efficacy and safety results from a Phase III, randomized controlled trial comparing the safety and efficacy of briakinumab with etanercept and placebo in patients with moderate to severe chronic plaque psoriasis. Br J Dermatol 2011;165(3):661–8.

[9] Gordon K, Langley R, Gu Y, et al. Efficacy and safety results from a phase III, randomized controlled trial comparing two ABT-874 dosing regimens to placebo in patients with moderate to severe psoriasis. Poster presented at Nineteenth EADV Congress. 2010.

[10] Reich K, Langley R, Papp K, et al. Efficacy and safety of ABT-874 versus methotrexate in patients with moderate to severe psoriasis. Poster presented at Nineteenth EADV Congress. 2010.

[11] Griffiths CE, Girolomoni G. Does p40-targeted therapy represent a significant evolution in the management of plaque psoriasis? J Eur Acad Dermatol Venereol 2012;26(Suppl. 5):2–8.

[12] Leonardi CL, Kimball AB, Papp KA, et al. Efficacy and safety of ustekinumab, a human interleukin-12/23 monoclonal antibody, in patients with psoriasis: 76-week results from a randomised, double-blind, placebo-controlled trial (PHOENIX 1). Lancet 2008;371(9625):1665–74.

[13] Papp KA, Langley RG, Lebwohl M, et al. Efficacy and safety of ustekinumab, a human interleukin-12/23 monoclonal antibody, in patients with psoriasis: 52-week results from a randomised, double-blind, placebo-controlled trial (PHOENIX 2). Lancet 2008;371(9625):1675–84.

[14] Griffiths CE, Strober BE, van de Kerkhof P, et al. Comparison of ustekinumab and etanercept for moderate-to-severe psoriasis. N Engl J Med 2010;362(2):118–28.

[15] Gottlieb AB, Cooper KD, McCormick TS, et al. A Phase I, double-blind, placebo-controlled study evaluating single subcutaneous administrations of a human interleukin-12/23 monoclonal antibody in subjects with plaque psoriasis. Curr Med Res Opin 2007;23(5):1081–92.

[16] Toichi E, Torres G, McCormick TS, et al. An anti-IL-12p40 antibody down-regulates type 1 cytokines, chemokines, and IL-12/IL-23 in psoriasis. J Immunol 2006;177(7):4917–26.

[17] Kauffman CL, Aria N, Toichi E, et al. A phase I study evaluating the safety, pharmacokinetics, and clinical response of a human IL-12 p40 antibody in subjects with plaque psoriasis. J Invest Dermatol 2004;123(6):1037–44.

[18] Krueger GG, Langley RG, Leonardi C, et al. A human interleukin-12/23 monoclonal antibody for the treatment of psoriasis. N Engl J Med 2007;356(6):580–92.

[19] González-Navajas JM, Lee J, David M, Raz E. Immunomodulatory functions of type I interferons. Nat Rev Immunol 2012;12(2):125–35. http://dx.doi.org/10.1038/nri3133.

[20] Dunn GP, Bruce AT, Sheehan KC, et al. A critical function for type I interferons in cancer immunoediting. Nat Immunol 2005;6(7):722–9.

[21] Crow MK. Type I interferon in organ-targeted autoimmune and inflammatory diseases. Arthritis Res Ther 2010;12(Suppl. 1):S5. http://dx.doi.org/10.1186/ar2886.

[22] Dunn GP, Bruce AT, Sheehan KC, et al. A critical function for type I interferons in cancer immunoediting. Nat Immunol 2005;6(7):722–9.

[23] Pestka S, Krause CD, Walter MR. Interferons, interferon-like cytokines, and their receptors. Immunol Rev 2004;202:8–32.

[24] Platanias LC. Mechanisms of type-I- and type-II-interferon-mediated signaling. Nat Rev Immunol 2005;5:375–86.

[25] van Boxel-Dezaire AHH, Rani MRS, Stark GR. Complex modulation of cell type-specific signaling in response to type i interferons. Immunity 2006;25:361–72.

[26] Baechler EC, Bauer JW, Slattery CA, et al. An interferon signature in the peripheral blood of dermatomyositis patients is associated with disease activity. Mol Med 2007;13(1–2):59–68.

[27] Bennett L, Palucka AK, Arce E, et al. Interferon and granulopoiesis signatures in systemic lupus erythematosus blood. J Exp Med 2003;197(6):711–23.

[28] Han GM, Chen SL, Shen N, et al. Analysis of gene expression profiles in human systemic lupus erythematosus using oligonucleotide microarray. Genes Immun 2003;4(3):177–86.

[29] Walsh RJ, Kong SW, Yao Y, et al. Type I interferon-inducible gene expression in blood is present and reflects disease activity in dermatomyositis and polymyositis. Arthritis Rheum 2007;56(11):3784–92.

[30] Higgs BW, Liu Z, White B, et al. Patients with systemic lupus erythematosus, myositis, rheumatoid arthritis and scleroderma share activation of a common type I interferon pathway. Ann Rheum Dis 2011;70(11):2029–36.

[31] Yao Y, Richman L, Higgs BW, et al. Neutralization of interferon-alpha/beta-inducible genes and downstream effect in a phase I trial of an anti-interferon-alpha monoclonal antibody in systemic lupus erythematosus. Arthritis Rheum 2009;60(6):1785–96.

[32] Merrill JT, Wallace DJ, Petri M, et al. Safety profile and clinical activity of sifalimumab, a fully human anti-interferon alpha monoclonal antibody, in systemic lupus erythematosus: a phase I, multicenter, double-blind randomised study. Ann Rheum Dis 2011;70(11):1905–13.

[33] McBride JM, Jiang J, Abbas AR, et al. Safety and pharmacodynamics of rontalizumab in patients with systemic lupus erythematosus: results of a phase I, placebo-controlled, double-blind, dose-escalation study. Arthritis Rheum 2012;64(11):3666–76.

[34] Higgs BW, Zhu W, Morehouse C, et al. Sifalimumab, an anti-IFN-α mAb, shows target suppression of a type I IFN signature in blood and muscle of dermatomyositis and polymyositis patients. Ann Rheum Dis 2013;February [Epub ahead of print].

[35] Liao AP, Salajegheh M, Nazareno R, Kagan JC, Jubin RG, Greenberg SA. Interferon β is associated with type 1 interferon-inducible gene expression in dermatomyositis. Ann Rheum Dis 2010;70(5):831–6.

[36] Cappelletti C, Baggi F, Zolezzi F, et al. Type I interferon and Toll-like receptor expression characterizes inflammatory myopathies. Neurology 2011;76(24):2079–88.

[37] Kotenko SV, Gallagher G, Baurin VV, et al. IFN-lambdas mediate antiviral protection through a distinct class II cytokine receptor complex. Nat Immunol 2003;4(1):69–77.

[38] Wang B, Higgs BW, Chang L, et al. Pharmacogenomic biomarker development and translational simulations to bridge clinical indications for an anti-interferon alpha receptor antibody. Clin Pharm Ther 2013;Feb. http://dx.doi.org/10.1038/clpt.2013.35.

[39] Abbas AK, Lichtman AH, Pillai S. B cell activation and antibody production. Cellular and Molecular Immunology. 6th ed. Philadelphia, PA: Saunders Elsevier; 2010. pp. 215–241.

[40] Reff ME, Carner K, Chambers KS, et al. Depletion of B cells *in vivo* by a chimeric mouse human monoclonal antibody to CD20. Blood 1994;83(2):435–45.

[41] Chynes RA, Towers TL, Presta LG. Inhibitory Fc receptors modulate *in vivo* cytotoxicity against tumor targets. Nat Med 2006;6:443–6.

[42] Owczarczyk K, Lal P, Abbas AR, et al. A plasmablast biomarker for non-response to antibody therapy to CD20 in rheumatoid arthritis. Sci Transl Med 2011;3(101):101ra92. http://dx.doi.org/10.1126/scitranslmed.3002432.

[43] De Vita S, Zaja F, Sacco S, De Candia A, Fanin R, Ferraccioli G. Efficacy of selective B cell blockade in the treatment of rheumatoid arthritis: evidence for a pathogenetic role of B cells. Arthritis Rheum 2002;46(8):2029–33.

[44] Edwards JC, Szczepanski L, Szechinski J, et al. Efficacy of B-cell-targeted therapy with rituximab in patients with rheumatoid arthritis. N Engl J Med 2004;350(25):2572–81.

[45] Cambridge G, Stohl W, Leandro MJ, Migone TS, Hilbert DM, Edwards JC. Circulating levels of B lymphocyte stimulator in patients with rheumatoid arthritis following rituximab treatment: relationships with B cell depletion, circulating antibodies, and clinical relapse. Arthritis Rheum 2006;54(3):723–32.

[46] Leandro MJ, Cambridge G, Ehrenstein MR, Edwards JC. Reconstitution of peripheral blood B cells after depletion with rituximab in patients with rheumatoid arthritis. Arthritis Rheum 2006;54(2):613–20.

[47] Roll P, Palanichamy A, Kneitz C, Dorner T, Tony HP. Regeneration of B cell subsets after transient B cell depletion using anti-CD20 antibodies in rheumatoid arthritis. Arthritis Rheum 2006;54(8):2377–86.

[48] Thurlings RM, Vos K, Wijbrandts CA, Zwinderman AH, Gerlag DM, Tak PP. Synovial tissue response to rituximab: mechanism of action and identification of biomarkers of response. Ann Rheum Dis 2008;67(7):917–25 [Epub 2007 October 26].

[49] Popa C, Leandro MJ, Cambridge G, Edwards JCW. Repeated B lymphocyte depletion with rituximab in rheumatoid arthritis over 7 yrs. Rheumatology 2007;46(4):626–30 [Epub 2006 December 19].

[50] Raterman HG, Vosslamber S, de Ridder S, et al. The interferon type I signature towards prediction of non-response to rituximab in rheumatoid arthritis patients. Arthritis Res Ther 2012;14(2):R95. http://dx.doi.org/10.1186/ar3819.

[51] Thurlings RM, Boumans M, Tekstra J, et al. The relationship between the type I interferon signature and the response to rituximab in rheumatoid arthritis. Arthritis Rheum 2010;62(12):3607–14. http://dx.doi.org/10.1002/art.27702.

[52] Jego G, Pascual V, Palucka AK, Banchereau J. Dendritic cells control B cell growth and differentiation. Curr Dir Autoimmun 2005;8:124–39.

[53] Vosslamber S, Raterman HG, van der Pouw Kraan CTM, et al. Pharmacological induction of IFN type I activity following therapy with rituximab determines clinical response in rheumatoid arthritis. Ann Rheum Dis 2011;70(6):1153–9. http://dx.doi.org/10.1136/ard.2010.147199 [Epub 2011 March 27].

[54] Verweij CL, Vosslamber S. New insight in the mechanism of action of rituximab: the interferon signature towards personalized medicine. Discov Med 2011;12(64):229–36.

[55] Merrill JT, Neuwelt CM, Wallace DJ, et al. Efficacy and safety of rituximab in moderately-to-severely active systemic lupus erythematosus: the randomized, double-blind, Phase II/III systemic lupus erythematosus evaluation of rituximab trial. Arthritis Rheum 2010;62(1):222–33.http://dx.doi.org/10.1002/art.27233.

[56] Compston A, Coles A. Multiple sclerosis. Lancet 2002;359(9313):1221–31. Review. Erratum in: Lancet 2002 August 24;360(9333):648.

[57] Schwid SR, Panitch HS. Full results of the evidence of interferon dose-response-European North American comparative efficacy (EVIDENCE) study: a multicenter, randomized, assessor-blinded comparison of low-dose weekly versus high-dose, high-frequency interferon beta-1a for relapsing multiple sclerosis. Clin Ther 2007;29(9):2031–48.

[58] Rudick RA, Lee JC, Simon J, Ransohoff RM, Fisher E. Defining interferon beta response status in multiple sclerosis patients. Ann Neurol 2004;56(4):548–55.

[59] Reske D, Walser A, Haupt WF, Petereit HF. Long-term persisting interferon beta-1b neutralizing antibodies after discontinuation of treatment. Acta Neurol Scand 2004;109(1):66–70.

[60] Rice GP, Paszner B, Oger J, Lesaux J, Paty D, Ebers G. The evolution of neutralizing antibodies in multiple sclerosis patients treated with interferon beta-1b. Neurology 1999;52(6):1277–9.

[61] van Baarsen LGM, Van der Pouw Kraan CTM, Kragt JJ, et al. A subtype of multiple sclerosis defined by an activated immune defense program. Genes Immun 2006;7(6):522–31 [Epub 2006 July 13].

[62] van Baarsen LG, Vosslamber S, Tijssen M, Baggen JM, van de, Voort E, et al. Pharmacogenomics of interferon-beta therapy in multiple sclerosis: baseline IFN signature determines pharmacological differences between patients. PLoS ONE 2008;3(4):e1927. http://dx.doi.org/10.1371/journal.pone.0001927.

[63] Comabella M, Lunemann JD, Rio J, et al. A type I interferon signature in monocytes is associated with poor response to interferon-β in multiple sclerosis. Brain Dec;132(Pt 12):3353–65. http://dx.doi.org/10.1093/brain/awp228.

[64] Bustamante MF, Fissolo N, Río J, et al. Implication of the Toll-like receptor 4 pathway in the response to interferon-β in multiple sclerosis. Ann Neurol 2011;70(4):634–45. http://dx.doi.org/10.1002/ana.22511.

[65] Hundeshagen A, Hecker M, Paap BK, et al. Elevated type I interferon-like activity in a subset of multiple sclerosis patients: molecular basis and clinical relevance. J Neurol 2012;9:140. http://dx.doi.org/10.1186/1742-2094-9-140.

[66] Zula JA, Green HC, Ransohoff RM, Rudick RA, Stark GR, van Boxel-Dezaire AH. The role of cell type-specific responses in IFN-β therapy of multiple sclerosis. Proc Natl Acad Sci USA 2011;108(49):19689–19694. http://dx.doi.org/10.1073/pnas.1117347108.

[67] Verweij CL, Vosslamber S. Relevance of the type I interferon signature in multiple sclerosis towards a personalized medicine approach for interferon-β therapy. Discov Med 2013;15(80):51–60.

[68] Rudick RA, Rani MR, Xu Y, et al. Excessive biologic response to IFNβ is associated with poor treatment response in patients with multiple sclerosis. PLoS One 2011;6(5):e19262. http://dx.doi.org/10.1371/journal.pone.0019262. [Epub 2011 May 13].

[69] Axtell RC, de Jong BA, Boniface K, et al. T helper type 1 and 17 cells determine efficacy of interferon-beta in multiple sclerosis and experimental encephalomyelitis. Nat Med 2010;16(4):406–12. http://dx.doi.org/10.1038/nm.2110. [Epub 2010 March 28].

[70] Bushnell SE, Zhao Z, Stebbins CC, et al. Serum IL-17F does not predict poor response to IM IFN-β-1a in relapsing-remitting MS. Neurology 2012;79(6):531–7.

[71] Barnes BJ, Kellum MJ, Pinder KE, Frisancho JA, Pitha PM. Interferon regulatory factor 5, a novel mediator of cell cycle arrest and death. Cancer Res 2003;63(19):6424–31.

[72] Schoenemeyer A, Barnes BJ, Mancl ME, et al. The interferon regulatory factor, IRF5, is a central mediator of toll-like receptor signaling. J Biol Chem 2009;284:2767–77.

[73] Hu G, Barnes BJ. IRF5 is a mediator of the death receptor-induced apoptotic signaling pathway. J Biol Chem 2009;284:2767–77.

[74] Graham RR, Kozyrev SV, Baechler EC, et al. A common haplotype of interferon regulatory factor 5 (IRF5) regulates splicing and expression and is associated with increased risk of systemic lupus erythemathosus. Nat Genet 2006;38:550–5.

[75] Pascual V, Farkas L, Banchereau J. Systemic lupus erythematosus: all roads lead to type I interferons. Curr Opin Immunol 2006;18:676–82.

[76] Rullo OJ, Woo JM, Wu H, et al. Association of IRF5 polymorphisms with activation of the interferon-α pathway. Ann Rheum Dis 2010;69:611–17.

Molecular Heterogeneity, Biomarker Discovery, and Targeted Therapy in Asthma

Joseph R. Arron, Jeffrey M. Harris

Genentech, Inc., South San Francisco, California

4.1 INTRODUCTION: ASTHMA HETEROGENEITY

Asthma has been described as a heterogeneous disorder on multiple clinical levels, including family history, age of onset, sex, body habitus, environmental triggers, symptoms, comorbidities, and responsiveness to specific interventions [1–6]. While in a broad sense some of these features align across populations of patients, there is no consensus on clinically defined asthma phenotypes. The concept of 'endotypes' in asthma has recently gained traction [7]: endotyping refers to an attempt to define specific subtypes of a disorder in terms of pathophysiologic mechanisms that may be amenable to targeted therapeutic interventions. However, to prove that a putative pathophysiologic mechanism 'drives' disease in a subset of asthma patients, successful intervention with therapeutics targeting that mechanism must be demonstrated.

While it has long been recognized that eosinophilic airway inflammation is a feature of asthma, a large and growing body of more recent evidence has shown that not all asthma patients have eosinophilic airway inflammation, and that asthma patients with variable levels of eosinophilia respond differently to anti-inflammatory therapies [3,6–13]. In particular, inhaled corticosteroids (ICS), the standard of care for most asthma patients, do not confer benefit to all asthmatics [14]. In several studies, ICS show greater efficacy in patients defined as having 'eosinophilic' disease than in 'noneosinophilic' patients. ICS treatment generally reduces the amount of airway inflammation as measured by sputum or tissue eosinophils or Fractional exhaled nitric oxide (FeNO) [11,15–17]. Furthermore, in ICS-responsive patients, titrating steroid dose to the level of airway eosinophilia or FeNO has been shown to reduce the rate of exacerbations and/or improve asthma control more effectively than

Y. Yao, B. Jallal, K. Ranade (Eds): Genomic Biomarkers for Pharmaceutical Development.
DOI: http://dx.doi.org/10.1016/B978-0-12-397336-8.00004-5

clinical guideline-based ICS dosing [18–20]. However, a subset of approximately 10% of asthma patients does not achieve adequate disease control even on high doses of ICS; this 'ICS-refractory' subset consumes a disproportionate share of health care expenditure and represents the most significant unmet medical need in asthma [21,22]. There are several non-mutually exclusive reasons why a given asthma patient may not achieve adequate disease control on ICS. These include poor adherence to prescribed therapy [17,23], intrinsic or acquired resistance to the molecular mechanism of action of steroids [24,25], or pathophysiology that is not inherently responsive to the effects of steroids. Development of new, molecularly targeted asthma therapies has been directed at the ICS-refractory population [26–28], but this population exhibits substantial heterogeneity [6], [29,30]. Thus, to interrogate whether a particular therapeutic candidate has the potential to show efficacy, it is important to assess the intervention in the patients most likely to have activity of the targeted pathway.

In this chapter we will describe our efforts, in collaboration with numerous colleagues in academia and industry, to define asthma heterogeneity in molecular terms and use that information to develop biomarkers that define asthma subsets and a novel therapy targeted at a subset of asthma patients. We will consider our findings in the context of recent publications of proof-of-concept clinical studies of candidate asthma therapeutics and an emerging appreciation of the challenges inherent in developing companion diagnostic tests alongside novel therapeutics. This is a rapidly evolving field and as of this writing, the therapeutic candidates and biomarkers discussed here had not yet been validated in pivotal clinical trials nor approved for the treatment of asthma. It should also be acknowledged that the issues and recommendations we raise with respect to the development of personalized therapies for asthma reflect our own experience and opinions and have not been formally endorsed by any regulatory authorities for any specific asthma therapies or biomarkers.

4.2 MOLECULAR HETEROGENEITY IN ASTHMA

4.2.1 Gene Expression Studies in the Asthmatic Airway

Although many studies have used genomic technologies to identify patterns that differentiate a given disease from other diseases or healthy controls, relatively few studies have used genomic technologies to characterize heterogeneity within a clinically defined non-neoplastic disorder such as asthma. One of the first studies to examine gene expression in airway tissues from asthma patients was conducted by Woodruff et al. [31], in which bronchial epithelial brushings from 42 mild-moderate asthma patients and 28 non-asthmatic controls were assessed by genome-wide expression microarrays. In this study, gene expression was compared between healthy controls and all asthmatics taken together. Using a fairly conservative analytical approach with Bonferroni correction for multiple testing, only 22 probesets achieved genome-wide significance for being significantly differentially expressed between asthma and control. Thirteen were expressed at higher levels and nine were expressed at lower levels in asthmatics than in control subjects. Given the considerable clinical and physiological differences in airway function between the asthma patients and controls in this study, one might have expected a greater number of differentially expressed genes. There are two potential explanations for why a stronger differential expression signal was not observed:

1. Bronchial epithelium itself is not a locus of molecular pathology in asthma.
2. Heterogeneity of gene expression within the group of asthmatics examined led to an underestimation of the gene expression differences when asthmatics, taken together, were compared to controls.

The former explanation is unlikely, as many studies have shown dysregulation of the bronchial epithelium in asthma, implicating defects in bronchial epithelium as an initiating factor in asthmatic airway inflammation [32]. The second explanation, given the known heterogeneity of infiltrating inflammatory cells in asthma, is more likely, as gene expression patterns present in some asthma patients and not others may have a mutually cancelling effect in the comparison examined (i.e., all asthmatics vs. all controls).

Among the significantly differentially expressed genes in asthmatic bronchial epithelial samples were three genes that had previously been shown to be responsive to IL13: *chloride channel, calcium-activated, family member 1* (CLCA1); *serine peptidase inhibitor, clade B, member 2* (serpinB2); and *periostin*. CLCA1 is an extracellular protein associated with chloride channels that mediates IL13-mediated differentiation of ciliated bronchial epithelial cells into mucus-producing goblet cells, a process known as mucous metaplasia [33–35]. SerpinB2, also known as plasminogen activator inhibitor-2, is a serine protease inhibitor that may be a negative regulator of Th1 responses in the airway [36,37]. Periostin is a matricellular protein and integrin ligand associated with fibrosis and remodeling that is inducible in stromal cells by various stimuli including IL13 and TGF-β [38–42].

As these three genes could be induced by IL13 in bronchial epithelial cells *in vitro* and IL13 had been implicated as a potential mediator of asthmatic airway inflammation and hence was under investigation as a therapeutic target, we hypothesized that bronchial epithelial gene expression patterns might be used to identify a subset of asthma patients more likely to have elevated activity of IL13 in their airways. Importantly, IL13 itself does not appear to be expressed by stromal cells such as bronchial epithelial cells [43], therefore, by serving as signal amplifiers, IL13-responsive genes may be better indicators of IL13 activity than the cytokine itself. To test this hypothesis, we examined whether CLCA1, serpinB2, and periostin were co-regulated with each other within samples from the subjects in the study. We found that each of the three genes was expressed at elevated levels in some but not all asthmatics, with substantial overlap between the asthmatic and control populations at lower expression levels. Furthermore, the expression of the three genes was highly intercorrelated in individual samples; i.e., if one gene was expressed at a given level in an individual subject within the range of values observed in the population, the other two genes were expressed at proportionally similar levels in that subject relative to the population. Based on these observations, we performed hierarchical clustering of all the asthma patients and controls in the study based on the expression levels of CLCA1, serpinB2, and periostin and found that the population divided into two distinct clusters: one cluster had high expression of all three genes, whereas the second cluster had low expression of all three genes. Importantly, these clusters did not separate by diagnosis: the cluster with high expression comprised about half of the asthma patients in the study, while the cluster with low expression comprised the other half of asthma patients interspersed with all of the non-asthmatic controls. Thus, on the basis of the bronchial epithelial expression levels of three genes, two molecularly distinct subsets of asthma patients emerged, one of which was, on the basis of expression of these genes, indistinguishable from healthy control subjects, despite having clear clinical manifestations of asthma [13].

To determine whether the coordinate elevated expression of CLCA1, serpinB2, and periostin was potentially due to the activity of IL13, we assessed the expression of type 2 cytokines by quantitative RT-PCR (qPCR) in endobronchial biopsies obtained contemporaneously with the epithelial brushings. Endobronchial biopsies are collected with forceps and sample multiple layers of bronchial mucosal tissue, including the epithelium as well as subepithelial layers including the lamina propria and airway smooth muscle bundles. In some asthma patients, the epithelium and lamina propria exhibit infiltration of inflammatory cells such as lymphocytes, eosinophils, neutrophils, mast cells, macrophages, and other cell types capable of producing inflammatory cytokines such as IL13. We found that, in endobronchial biopsies from patients with high bronchial epithelial expression of CLCA1, serpinB2, and periostin, there was significantly elevated expression of IL13 and IL5 compared to patients with low bronchial epithelial expression of CLCA1, serpinB2, and periostin [13]. Since IL5 and IL13 were defined at the time as cytokines typically expressed in T-helper type 2 (Th2) polarized CD4+ T cells, we called asthma patients in the cluster with high expression of the three genes 'Th2-high' and patients in the cluster with low expression of the three genes 'Th2-low'.

While the gene expression patterns clearly identified distinct subsets of asthma patients, the cross-sectional nature of the analysis could not rule out the possibility that the gene expression observed was transient in response to environmental stimuli (e.g., aeroallergen exposures) near the time of bronchoscopy. To determine whether the gene expression patterns bore any relationship to independent measures of asthma, we compared the 'Th2-high' and 'Th2-low' asthma patients on the basis of multiple other clinical and pathological assessments. Both subsets of asthma patients had significant airway obstruction as measured by spirometry (forced expiratory volume in one second, FEV1) and reversibility to β_2-adrenergic agonists compared to healthy controls, and while the 'Th2-high' subset was sensitive to lower provocative concentrations of methacholine (a measure of airway hyperresponsiveness, AHR) than the 'Th2-low' subset of asthma patients, both asthma groups were significantly more sensitive to methacholine than healthy controls. However, the two asthmatic subsets differed significantly on other measures, with the 'Th2-high' asthmatics exhibiting elevated eosinophils in bronchoalveolar lavage (BAL) and peripheral blood, altered mucus composition, and thicker airway reticular basement membranes (a measure of bronchial fibrosis) compared to 'Th2-low' asthmatics. Thus, while both groups of asthmatics met a clinical definition of asthma (reversible airway obstruction and AHR), they were distinct with respect to other asthma pathologies. Upon repeat bronchoscopy one week after the initial bronchoscopy, 'Th2-high' and 'Th2-low' asthmatics continued to have similar gene expression patterns, suggesting short-term temporal stability of the phenotype. Importantly, when randomized to an eight week course of ICS or placebo, only the 'Th2-high' asthmatics demonstrated significant improvements in lung function on ICS, while the 'Th2-low' asthmatics did not exhibit any FEV1 changes relative to placebo-treated patients; this FEV1

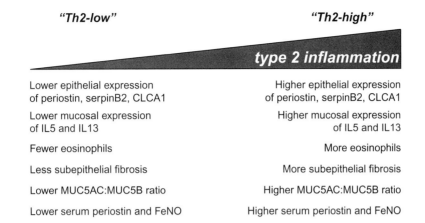

"Th2-low"	"Th2-high"
Lower epithelial expression of periostin, serpinB2, CLCA1	Higher epithelial expression of periostin, serpinB2, CLCA1
Lower mucosal expression of IL5 and IL13	Higher mucosal expression of IL5 and IL13
Fewer eosinophils	More eosinophils
Less subepithelial fibrosis	More subepithelial fibrosis
Lower MUC5AC:MUC5B ratio	Higher MUC5AC:MUC5B ratio
Lower serum periostin and FeNO	Higher serum periostin and FeNO

FIGURE 4.1 Asthma patients exhibit a continuum of type 2 airway inflammation. While all asthma patients have reversible airway obstruction and airway hyper-responsiveness, there is a variable level of type 2 airway inflammation as defined by gene expression in the bronchial mucosa. The degree of gene expression related to type 2 inflammation is correlated to pathophysiological features of asthma such as eosinophilia, fibrosis, mucus composition, and levels of non-invasive biomarkers such as serum periostin and FeNO.

improvement was reversed within one week of cessation of ICS treatment [13]. Taken together, these corroborative data suggest that the molecularly defined phenotypes are relatively stable and clinically meaningful in mild-moderate asthma.

Using the bronchial epithelial gene expression designation as a discriminating factor, we assessed gene expression in matched endobronchial biopsy tissue from a subset of asthma patients and controls in the study (N = 27 and 13, respectively). Despite the smaller number of subjects and greater variability of sample composition in biopsies as opposed to epithelial brushings, by comparing 'Th2-high' asthmatics to 'Th2-low' asthmatics and healthy controls (rather than 'asthma' vs. 'control'), we identified 93 differentially expressed probes corresponding to 79 genes using similarly conservative analytical methods [44]. This broader set of highly co-regulated genes in bronchial mucosa enabled a more granular quantitative description of the molecular phenotype, which could be condensed to a single gene expression score. This 'Th2 signature' score varied continuously across the dataset and correlated continuously with independent measures of type 2 inflammation including blood and airway eosinophil levels. In addition, the 'Th2 signature' was highly positively correlated with the expression levels of the Th2 cytokines IL13 and IL5, as well as with eosinophil chemoattractants CCL26 and CCL13, while it was negatively correlated with the expression level of the Th1 cytokine IL12A. This molecular definition of asthma heterogeneity along a continuum of type 2 inflammation appeared to correspond well with the cytologic definition of asthma heterogeneity according to airway eosinophilia. The greater precision afforded by the biopsy gene expression signature underscored the reality that while it is often a convenient shorthand to define asthma phenotypes as discrete entities, there is clearly a continuum of type 2 inflammation across the population. The findings described in [13] and [44] are summarized conceptually in Fig. 4.1.

4.2.2 Translation to Non-Invasive Biomarkers

While these initial findings were encouraging, they raised two issues for therapeutic development:

1. These studies were conducted in mild-moderate asthmatics not taking ICS, but the major unmet medical need in asthma is in more severe patients who are inadequately controlled despite ICS treatment.
2. Although precise molecular phenotyping is possible using airway samples, collecting airway samples requires bronchoscopy or sputum induction, which is technically challenging, time and labor intensive, not widely available outside specialty pulmonary practices, and thus logistically impractical in large-scale multi-center clinical trials or in primary care settings, where most asthma patients are managed.

To address these issues, we conducted a new observational study ('BOBCAT') in 67 moderate-severe asthma patients who remained poorly controlled despite high-dose ICS treatment, in which we collected matched sets of induced sputum, endobronchial biopsies, and peripheral blood. The objectives of BOBCAT were:

1. to characterize the extent and variability of eosinophilic airway inflammation in moderate-severe asthma;
2. to identify non-invasive biomarkers of eosinophilic airway inflammation for use in interventional trials of investigational therapeutic candidates targeting type 2 inflammation.

Consistent with previous reports [1,2] and [45], we observed a range of eosinophilic airway inflammation in the BOBCAT cohort, with some but not all patients exhibiting dramatically elevated eosinophils in sputum and/or bronchial mucosal tissue [46]. Having confirmed a range of airway eosinophilia in moderate-severe asthma that remained symptomatic despite high-dose ICS treatment, we sought to identify non-invasive biomarkers of the eosinophilic airway phenotype.

Th2 cytokine proteins such as IL13 are present at very low levels in peripheral blood, below the limit of detection of all but the most sensitive assays. A recent report using such an assay showed that circulating levels of IL13 protein were around 1 pg/ml and were not appreciably different between asthma patients and healthy controls [47]. This concentration of IL13 is 10^3–10^4-fold lower than the amount of recombinant cytokine necessary to activate signaling in target cells *in vitro* [48]. Given that IL13 receptors are highly expressed in stromal cells in the airways, it is likely that, in patients with asthma, IL13 is produced in the airways at the site of inflammation where the majority binds to and/or is consumed by stromal cells in the immediate vicinity. Thus the level of IL13 protein in peripheral blood is unlikely to be a robust systemic biomarker of its own activity in the airways. We hypothesized that IL13 might induce the expression of genes in bronchial mucosa that could be translated to non-invasive biomarkers of the 'Th2-high' molecular phenotype in the airways. We will next consider three different examples of biomarkers identified through this process: serum periostin, FeNO, and blood eosinophil counts.

We searched among the genes correlated with the Th2 signature in epithelial brushings and endobronchial biopsies for those encoding extracellular proteins to determine whether those proteins could be detected in peripheral blood. For example, CLCA1 is expressed on the apical aspect of bronchial epithelial cells near the airway lumen [49], consistent with its proposed role in mucus secretion [35]. Its expression is therefore likely to be densest at the luminal surface of the bronchial mucosa and CLCA1 protein, if shed in soluble form, is most likely detectable in samples taken from the airway such as induced sputum or BAL [34]. Periostin, on the other hand, is secreted from the basolateral aspect of bronchial epithelial cells in response to IL13 stimulation [40] and may also be expressed in subepithelial fibroblasts in the lamina propria [39], more proximal to blood vessels serving the bronchial mucosa. For this reason, we hypothesized that periostin may be more likely to be detectable in peripheral blood than CLCA1. In addition, some reports have described relatively high levels of circulating periostin protein (10s–100s ng/ml) in patients with epithelially-derived cancers [50–53], suggesting that substantial levels of the protein might be detectable in peripheral blood. Thus we prioritized periostin for further investigation, including assay development. Many other soluble blood protein candidates were evaluated along similar lines but will not be discussed further here. Nitric oxide synthase 2 (NOS2), encoding inducible nitric oxide synthase (iNOS), was among the genes most significantly associated with the Th2 signature in both bronchial epithelial brushings and endobronchial biopsies [44]. iNOS is an enzyme that catalyzes the production of nitric oxide (NO) from L-arginine, and its expression can be induced in bronchial epithelial cells by IL13. NO is an exhaled gas that is detectable in exhaled air, and FeNO has been used extensively in the diagnosis and management of asthma and clinically validated instruments for its detection are in wide use [54,55]. Accordingly, we advanced FeNO as another candidate marker of airway IL13 activity. The Th2 signature is highly correlated with the expression of IL5, a Th2 cytokine that is an obligate factor for eosinophil hematopoiesis, activation, and survival [56]. IL13 induces the expression of multiple CCR3-binding chemokines such as CCL11, CCL13, CCL24, and CCL26, which serve as eosinophil chemoattractants [57–59]. Local expression of IL5 and IL13 in asthmatic airway tissue can contribute to airway eosinophilia via expansion and recruitment of eosinophils and might contribute to increased trafficking of eosinophils through the systemic circulation. Therefore, we advanced peripheral blood eosinophil counts as a third candidate marker of airway IL13 activity. Each of these three markers (serum periostin, FeNO, and blood eosinophils) is mechanistically linked to the activity of Th2 cytokines in the airways and could be measured noninvasively. The relationships between IL5, IL13, bronchial mucosa, and the markers discussed here are depicted schematically in Fig. 4.2.

FeNO could be detected using existing instruments [55], and peripheral blood eosinophil counts could be derived from a complete blood count (CBC) with differential, which is a widely available clinical laboratory assay. However, there were no validated assays for peripheral blood periostin. 'Research-grade' immunoassays often use polyclonal antibodies for either capture or detection in a sandwich enzyme-linked immunosorbent assay (ELISA) format. However, the use of polyclonal antibodies introduced at least two challenges for the development of a serum periostin assay, one of which is specific to periostin and one of which is a general challenge for immunoassay-based clinical diagnostic tests. Periostin protein exists in multiple alternatively spliced variants [60], and several of these variants are expressed simultaneously in bronchial epithelial brushings from asthma patients (G. Jia and J. R. Arron, unpublished observations). While exons 1–16 are conserved in most expressed variants, exons 17–23 undergo alternative splicing. Thus certain epitopes from the C-terminal portion of periostin

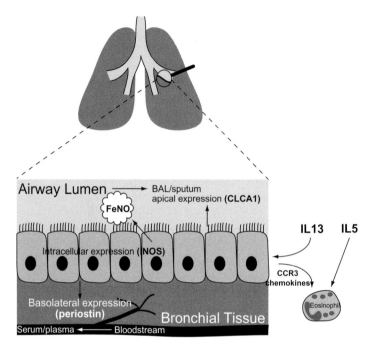

FIGURE 4.2   Relationships between type 2 cytokines, bronchial mucosa, and biomarkers. IL13 induces the expression of genes encoding periostin, CLCA1, iNOS, and CCR3-binding chemokines (e.g., CCL13, CCL26) in bronchial epithelial cells. IL5 induces eosinophil hematopoiesis and CCR3-binding chemokines recruit eosinophils to bronchial tissue. In bronchial epithelial cells, CLCA1 protein is expressed on the apical surface, iNOS is expressed intracellularly, and periostin is secreted from the basolateral surface. Secreted periostin protein is detectable in peripheral blood, while exhaled nitric oxide (FeNO) is detectable in exhaled breath.

may be variably present in a given clinical sample. Polyclonal antibodies raised to full-length periostin protein may contain mixtures of individual antibodies that bind to epitopes present in only a subset of periostin splice variants. To develop a prototype clinical-grade assay for periostin, we selected monoclonal antibodies for both capture and detection that bound to epitopes in the conserved N-terminal region of the known splice variants of periostin, so that the assay provided a true estimate of total periostin protein in a clinical sample [46]. Figure 4.3 depicts a simplified theoretical example of how varying isoform mixtures in two clinical samples with identical levels of total periostin protein could yield different outcomes with a polyclonal antibody (pAb)-based immunoassay. A more general challenge in developing regulated clinical diagnostic tests is reproducibility of the assay over time across manufacturing lots. Production of pAbs relies on affinity purification of antibodies from serum of immunized animals, which is an exhaustible supply of reagent. Once the serum from a particular bleed is consumed, additional bleeds from the same animal taken at different times and/or serum from other immunized animals must be used to produce additional pAbs, which are likely to consist of different mixtures of individual antibodies over time, making assay standardization difficult. As monoclonal antibodies (mAbs) are produced from cell lines and consist of single purified clones, it is substantially more feasible to ensure assay consistency over time across manufacturing lots.

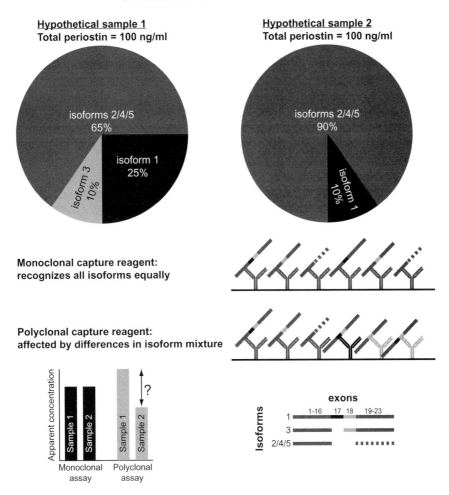

FIGURE 4.3 Potential differences between monoclonal and polyclonal periostin immunoassays. The C-terminal portion of periostin protein is subject to alternative splicing in exons 17–23 (lower right). Two hypothetical samples with identical total periostin levels but with differing isoform composition (top) will yield similar results with a monoclonal antibody-based assay that recognizes epitopes conserved in all splice variants but could yield different results using a polyclonal antibody-based assay that contains a mixture of antibodies, some of which recognize epitopes unique to certain splice variants.

Using the intensive airway sampling in BOBCAT to define phenotypes according to the degree of sputum and/or tissue eosinophilia, we evaluated the relationships between airway eosinophilia and FeNO, blood eosinophils, and serum periostin (using the mAb-based prototype periostin ELISA described above). Consistent with previous reports, we found that measures of sputum and tissue eosinophils were only weakly positively correlated. While some patients had elevations of either sputum or tissue eosinophils, others had elevated eosinophils in neither or both compartments. Using either discrete classifiers with pre-specified cutoff values for sputum and tissue eosinophils or continuous measures, we found that serum periostin, FeNO, and to a lesser extent, blood eosinophils increased across the population with increasing levels of airway eosinophilia. Of these three biomarkers, serum periostin predicted airway eosinophilia with greater sensitivity and specificity than either FeNO or blood eosinophils and exhibited less intrapatient variability over time [46], although as discussed above, the three measures reflect similar biological processes and thus it was unsurprising that they were intercorrelated, if weakly, across the population.

4.3 BIOMARKER-GUIDED PATIENT SELECTION IN THERAPEUTIC TRIALS

Several investigational therapies targeting type 2 inflammation in asthma have undergone early clinical development. Detailed discussions and listings of these agents can be found elsewhere [26–28]; we will focus here on proof-of-concept studies that have used biomarkers to select patients most likely to demonstrate clinical benefit from agents targeting IL5 and IL13, as agents targeting these cytokines have the largest amount of publicly available clinical trial data.

4.3.1 Agents Targeting IL5

As discussed above, IL5 is an obligate hematopoietic factor for eosinophils, regulating their differentiation, activation and survival. Three humanized monoclonal antibody therapies directed at IL5 or its receptor have been evaluated in randomized, placebo-controlled trials in asthma patients. The first agent to be described was mepolizumab (anti-IL5), which failed to demonstrate clinical benefit in an unselected population of moderate asthma patients incompletely controlled on ICS therapy, despite showing significant pharmacodynamic effects on blood and sputum eosinophils [61]. Three possible explanations for the lack of observed clinical benefit despite evidence for a biological effect in the study are:

1. Incomplete biological effect on eosinophils in bronchial mucosal tissue
2. Selection of a primary outcome measure – in this case, the asthma control questionnaire (ACQ) [62] – that did not accurately reflect the relationship of the biological target to the clinical presentation of asthma
3. Inadequate selection of those patients whose disease is likely to be associated with eosinophil activity.

A non-significant trend toward reduction in asthma exacerbation rates (a secondary outcome measure) was observed in the study, although there were not enough exacerbation events during the period of observation to have sufficient statistical power to adequately evaluate this outcome measure [61]. Two subsequent studies addressed the issues of outcome and patient selection by enrolling only severe asthma patients with greater than 3% sputum eosinophils who had experienced multiple exacerbations in the prior year [63,64], as past exacerbation history is a prognostic indicator of future exacerbations [65]. Each of these studies demonstrated significant clinical benefit of IL5 blockade in terms of asthma exacerbation reduction. Reslizumab, another anti-IL5, was investigated in moderate-severe patients selected on the basis of $\geq 3\%$ sputum eosinophils, and showed significant benefit in terms of lung function, with trends toward effects on ACQ and exacerbation rates [66]. Interestingly, patients in that study with concomitant nasal polyposis, a common comorbidity of eosinophilic asthma particularly evident in patients with aspirin-intolerant asthma, exhibited the greatest magnitude of clinical benefit. A third agent targeting IL5 is benralizumab, an antibody directed against IL5Rα that is cytotoxic to eosinophils. In a single dose study in severe eosinophilic asthmatics presenting with exacerbations, the rate of exacerbations in the subsequent 24 weeks was significantly reduced by benralizumab treatment [67]. Taken together, these relatively small proof-of-concept studies demonstrate that targeting IL5 is capable of partially reducing the rate of exacerbations in asthma patients that have evidence of eosinophilic airway inflammation and a past history of asthma exacerbations.

To build on the small proof-of-concept studies showing that IL5 blockade could potentially reduce the rate of exacerbations in moderate-severe asthmatics with evidence of eosinophilic airway inflammation, a larger study ('DREAM') was conducted, in which 616 patients were randomized (1:1:1:1) to receive placebo or one of three dose levels of mepolizumab for 52 weeks [68]. The primary outcome measure was the rate of clinically significant asthma exacerbations during the treatment period. For study entry, patients were required to have had a history of at least two exacerbations in the previous year and evidence of eosinophilic airway inflammation by fitting at least one of the following criteria: sputum eosinophil count $\geq 3\%$, FeNO ≥ 50 ppb, blood eosinophils $\geq 0.3 \times 10^9$/L, or 'prompt deterioration of asthma control after a 25% or less reduction in regular maintenance inhaled or oral corticosteroids'. With this study design, statistically significant reductions in exacerbation rates were observed in all dose arms, with the exacerbation rate in treated patients reduced by 48%, 39%, and 52% in the three active arms relative to the placebo arm with no apparent dose ranging effect. While blood and sputum eosinophils were significantly reduced in the mepolizumab-treated patients as compared to placebo-treated patients, there were no significant effects on FeNO, FEV1, or ACQ in this study. Overall, the DREAM study provided further compelling evidence that IL5 blockade may significantly reduce asthma exacerbations provided that patients with evidence of eosinophilic airway inflammation and a recent history of exacerbations are selected.

However, it remains unclear how the patient selection criteria (i.e., either sputum or blood eosinophilia, or elevated FeNO, or deterioration of control upon steroid reduction) will be translated into practicable guidelines from the standpoints of regulatory approval and broad clinical applicability. In particular, no single study has been described in which patients predicted to benefit from IL5 blockade (i.e., patients with elevated sputum or blood eosinophils or FeNO) were compared directly to patients predicted not to benefit (i.e., patients below the threshold for all three biomarkers). While it may be reasonable to assume that IL5 blockade might benefit patients with 'eosinophilic asthma', however it may be defined, in order to prove this hypothesis it is equally important to demonstrate that the 'diagnostic-negative' population *fails* to show benefit as it is to demonstrate that the 'diagnostic-positive' population does show benefit (provided that exposure to drug is deemed not to pose significant safety risks to the 'diagnostic-negative' population). A corollary of this point is the definition of a suitable cutoff for the diagnostic biomarker: while 3% sputum eosinophils, 0.3×10^9 blood eosinophils/L, or 50 ppb FeNO may sound like reasonable cutoffs based on observational studies, it is unclear whether those cutoffs are optimal for selecting IL5 blocking treatment in individual patients.

This issue is further complicated by the imprecision of using severe exacerbations as an outcome measure. While a severe exacerbation is an objectively definable event, exacerbations are rare in individual patients ('frequent exacerbators' may have 2 per year; many patients in each arm of the DREAM trial had no exacerbations during the study). Thus to provide adequate statistical power to determine an optimal biomarker cutoff to predict benefit in terms of asthma exacerbations, a study may become prohibitively large, depending on what is meant by an 'optimal' cutoff, as any cutoff will necessarily reflect a compromise between the magnitude of treatment benefit and the size of the treatment-eligible population. Finally, given the temporal intrapatient variability of each of these biomarkers, a given patient is likely to oscillate between 'diagnostic-positive' and 'diagnostic-negative' states over time. These issues are not unique to IL5 blocking agents, as we shall describe in the next section.

4.3.2 Agents Targeting IL13

Four biologic agents targeting IL13 (±IL4) have been investigated in Phase II proof-of-concept clinical studies in moderate-severe asthmatics whose asthma is inadequately controlled despite ICS treatment: pitrakinra [69], AMG 317 [70], tralokinumab [71], and lebrikizumab [72]. Pitrakinra is an inhaled IL4 mutein that binds to, but does not signal through, IL4Rα, the shared receptor component for IL4 and IL13 signaling, thereby blocking access of IL4 and IL13 to the IL4Rα/γc and IL4Rα/IL13Rα1 receptor complexes. AMG 317 is a humanized monoclonal antibody that binds to IL4Rα, thereby blocking IL4 and IL13 signaling. Tralokinumab and lebrikizumab are humanized monoclonal antibodies that bind to and block signaling downstream of IL13. Pitrakinra was administered twice daily by inhalation while AMG 317 (weekly), tralokinumab (q2 weeks), and lebrikizumab (q4 weeks) were all administered by subcutaneous injection. The pitrakinra, AMG 317, and tralokinumab studies were dose ranging, with approximately equal distributions of patients across each dose and placebo arm while the lebrikizumab study tested a single dose level with a 1:1 distribution vs. placebo.

While each study differed in specific design features, all four studies assessed lung function via FEV1, symptom control via ACQ, and exacerbation rates over at least 12 weeks of treatment. Pitrakinra, AMG 317, and tralokinumab failed to show significant clinical benefit relative to placebo in terms of their primary endpoints (exacerbation rate for pitrakinra and ACQ for AMG 317 and tralokinumab) in all comers [69–71], while lebrikizumab demonstrated a statistically significant, albeit modest benefit in terms of its primary endpoint (FEV1) in all comers [72]. Evaluating each agent across common outcome measures, none of the four demonstrated significant benefits in terms of ACQ, which may reflect regression to the mean, as all four studies pre-specified poor asthma control (ACQ ≥ 1.5) at study entry and patients in the placebo arms exhibited substantial ACQ improvement in each study. In a *post hoc* analysis of the AMG 317 study, patients in the highest tertile of baseline ACQ scores in the highest dose arm trended toward FEV1 improvement [70]. FEV1 was a secondary endpoint in the tralokinumab study, and while there was a greater mean magnitude of FEV1 improvement at 13 weeks in the combined tralokinumab-treated arms vs. placebo, this effect failed to reach statistical significance (p = 0.072) although the effect was nominally significant (p = 0.041) in the highest dose arm. In a substudy of patients that provided induced sputum samples at baseline, there was greater FEV1 improvement in tralokinumab-treated patients with $\geq$10 pg/ml IL13 in sputum supernatants than in tralokinumab-treated patients with <10 pg/ml sputum IL13 or all placebo patients taken together. There was also a greater magnitude of FEV1 improvement observed in patients with baseline peripheral blood eosinophil counts $\geq 0.3 \times 10^9$/L than in patients with $< 0.3 \times 10^9$/L blood eosinophils [71]. There is no available published data at this time on the effect of pitrakinra on FEV1. However, while pitrakinra failed to significantly affect exacerbation rates vs. placebo in all comers, secondary analyses stratifying the population by baseline FeNO, blood eosinophil counts, or a single nucleotide polymorphism in IL4RA showed greater exacerbation rate reductions for the treated group in each defined subset, although there was little overlap between subsets as defined by each baseline characteristic [69,73]. Overall, each of these studies suggests that IL13 blockade may provide clinical benefit in terms of lung function and exacerbation reduction in some, but not all, moderate-severe asthma patients inadequately controlled on ICS.

In the lebrikizumab Phase II proof-of-concept study ('MILLY'), we stratified treatment assignments at randomization according to a composite of serum IgE and blood eosinophil counts [72], which effectively differentiated mild-moderate asthmatics not taking ICS according to bronchial epithelial Th2 signature status [74]. However, prior to study unblinding, we pre-specified a data analysis plan in which we stratified the outcome analyses according to baseline serum periostin level. This change in stratification was because the prototype periostin diagnostic assay was not available at the start of the study but was ready by the time all the patients had been enrolled in the study before unblinding, and we had confidence in its ability to predict airway eosinophilia in moderate-severe asthma patients based on the BOBCAT study [46]. As an exploratory analysis, we also evaluated the ability of baseline FeNO levels to predict clinical benefit. Because we had observed continuous distributions of these biomarkers, which scaled continuously with airway eosinophil measures in BOBCAT, there was no obvious biologically definable cutoff to pre-specify. Thus, to maximize statistical power by creating approximately equally sized subgroups, we pre-specified the median values of the biomarkers at baseline as cutoffs. It should be noted that the median FeNO level in this study (21 ppb) was considerably lower than the pre-specified FeNO cutoff in the mepolizumab DREAM study (50 ppb) [68]. Considering all comers, lebrikizumab demonstrated a significant FEV1 benefit vs. placebo at 12 weeks (5.5%, 95% CI 0.8–10.2%). Dichotomizing the population according to median baseline serum periostin or FeNO levels, we observed comparable magnitudes of enrichment for FEV1 improvement in the 'diagnostic-positive' subsets as compared to the 'diagnostic-negative' subsets (8.2% vs. 1.6% for periostin and 8.6% vs. 1.9% for FeNO, both with wide but non-overlapping confidence intervals). Importantly, this stratification demonstrated that a significant FEV1 improvement vs. placebo was observed in the 'diagnostic-positive' subsets but there was no statistically significant FEV1 benefit in the 'diagnostic-negative' subsets; thus the significant effect observed in all comers was driven primarily by the effect in the 'diagnostic-positive' subgroups. In assessing serum periostin and FeNO levels at two visits one week apart during the pre-dose run-in period, we observed considerably more intrapatient variability in FeNO than in periostin, with a coefficient of variation (CV) of 19.3% (95% CI 17.4–22.2%) for FeNO vs. 5.0% (95% CI 4.4–5.6%) for serum periostin [72]. This greater level of intrapatient variability in FeNO than in periostin was also observed across multiple visits in the BOBCAT study [46]. Blood eosinophils, although not prioritized as predictive biomarkers in the analyses performed, enriched for responsiveness to mepolizumab and tralokinumab as described above, in the MILLY study were comparable in intrapatient variability to FeNO with a CV of 21.3% (95% CI 18.7–24.0%) [72]. In a subsequent analysis, we found that while baseline serum periostin and FeNO levels were significantly positively correlated among patients in the trial, there were subsets of patients that were above the median for one, but not the other biomarker, while others were either below or above the median for both biomarkers. Of these four subsets, the one with both serum periostin and FeNO greater than the median exhibited the greatest magnitude of FEV1 improvement relative to placebo [75]. While the size and duration of the study were not adequately powered to show benefit in terms of severe exacerbations within the 20 week treatment period, pharmacodynamic effects of lebrikizumab were observed for as long as 32 weeks. Considering a 32 week observation period, there was a significant reduction in the rate of severe exacerbations in the lebrikizumab arm as compared to the placebo arm; while there was a greater magnitude of exacerbation reduction in 'diagnostic-positive' patients, this effect was not statistically significant [76].

Crucial mechanistic insights about the relationships between IL13 and non-invasive biomarkers came from pharmacodynamic analyses. Lebrikizumab-treated patients exhibited a small increase in peripheral blood eosinophil counts, which may be a consequence of decreased levels of IL13-induced eosinophil-attracting chemokines, such as CCL13 [72]. While it remains to be formally determined whether IL13 blockade reduces the levels of eosinophils in bronchial mucosal tissue, the increased blood eosinophil counts and decreased CCL13 levels are consistent with this hypothesis. Lebrikizumab treatment induced sustained reductions in the mean levels of FeNO to below 20ppb in the study, which is within the range of nonasthmatic healthy control patients [55]. Interestingly, lebrikizumab-treated patients with serum periostin levels above the median at baseline exhibited a significant decrease in serum periostin levels, while there was no significant change in serum periostin levels in patients whose baseline serum periostin levels were below the median [77]. Taken together, these pharmacodynamic effects in which FeNO and serum periostin are 'normalized' (but not reduced to undetectable levels) suggest that IL13 is a primary driver of excess FeNO in asthma patients and that the excess serum periostin levels observed in 'periostin-high' patients with asthma are due to the effects of IL13.

By assessing the effects of lebrikizumab in a stratified population comprising patients with greater and lesser degrees of activity of the targeted pathway, we have tested the hypothesis that 'diagnostic-positive' patients are more likely to exhibit clinical benefit from IL13 blockade than 'diagnostic-negative' patients. Given the size of the study, the width of the confidence intervals for the clinical outcome measures, and the intrapatient variability in FEV1 and predictive biomarkers, it is difficult from these data to more precisely define biomarker cutoffs to differentiate between patients most likely to show clinical benefit from IL13 blockade and those less likely to benefit. Clearly, these encouraging initial findings will need to be replicated and refined in larger cohorts.

4.4 COMPANION DIAGNOSTIC DEVELOPMENT

As detailed above, the tasks of discovering candidate biomarkers and devising prototype assays to detect those biomarkers early in the drug development process is difficult and complex. However, these basic and translational research questions can be pursued according to precedents set by earlier efforts. On the other hand, translating biomarker data from early clinical development studies into validated companion diagnostic tests is much less well-explored territory. In this section we will offer insights and opinions gained from our experiences but caution that this is very much work in progress with considerable uncertainty as to how the regulatory framework for companion diagnostics may evolve. We will provide references to existing regulatory guidance documents but acknowledge that there are presently few published data on the implementation of biomarker selection and companion diagnostic assay validation in clinical studies of candidate asthma therapeutics.

A primary challenge of companion diagnostic development is having enough data and confidence to invest the resources to develop these assays on an appropriate platform and with sufficient quality control in place well before the pivotal Phase III drug trials are initiated. This early investment in the assays for use in Phase III is because these trials will be used to register not only the drug, but also the specific way in which the diagnostic test is to be used in conjunction with the drug. Given the complexity of regulatory requirements for both drugs and diagnostics, companion diagnostic development dictates that the work begins to be at risk very early in the development process and that as many opportunities as possible are taken to generate data about how the biomarker performs. As such, the discussion here will focus on diagnostic biomarker development during early clinical development, up to the start of Phase III pivotal trials, by which time the diagnostic biomarker should be established if it is to be an essential component of registration.

Figure 4.4 depicts a schematic early development timeline for a candidate asthma drug and a companion diagnostic. Assuming suitable pilot data on the relationships between the biomarker, the therapeutic target, and the disease state are available, assay development with a partner diagnostic company may take several years. Thus, to allow sufficient time to develop a robust clinical biomarker assay prior to the start of pivotal Phase III studies, exploratory biomarker discovery must take place very early, preferably prior to the start of Phase I clinical studies. A technically validated assay capable of reliably reporting biomarker values in relevant clinical samples should be in place prior to unblinding Phase II proof-of-concept clinical studies to enable a prospective stratified analysis. The outcome of the proof-of-concept trial determines the strategy for diagnostic implementation in pivotal trials. Three potential outcomes are depicted:

1. The diagnostic enriches for clinical benefit in Phase II but the Phase II study is underpowered to identify a precise cutoff based on the outcome measures. Phase III should enroll all comers with a pre-specified analysis of patients stratified according to baseline biomarker levels. In this case, a provisional cutoff value may be pre-specified with the understanding that the greater statistical power afforded by larger numbers of patients enrolled in the Phase III study could enable post-hoc optimization of the biomarker cutoff. The regulatory pathway for this option remains to be defined, and any post-hoc adjustments would need to be based on relatively unambiguous data.
2. In the rare event that a precise diagnostic cutoff is identified in Phase II and there is no clear clinical benefit to 'diagnostic-negative' patients, Phase III could proceed with enrollment of only 'diagnostic-positive' patients. As discussed in this chapter, given the variability of approvable outcome measures and the continuous, rather than dichotomous, nature of asthma subtypes, we do not believe this scenario is likely for asthma therapies currently under investigation.

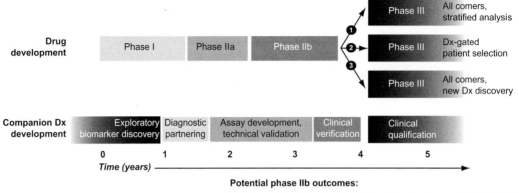

FIGURE 4.4 Hypothetical timelines for early clinical development of an asthma drug and a companion diagnostic (Dx). To prospectively validate a Dx test in a phase III pivotal trial, a candidate biomarker should be identified via exploratory studies while the drug candidate is in very early stages of clinical development. This early activity will allow selection of a partner diagnostic company and development and technical validation of a prototype clinical assay in time to permit clinical verification that the biomarker enriches for clinical benefit in a phase II proof-of-concept study without delaying the clinical development timeline. Because the companion Dx development is done at risk, there are several potential implications for how the biomarker may be used in a pivotal trial; three possibilities as discussed in the text are shown.

3. If the diagnostic hypothesis fails to be validated in the proof-of-concept study but there is evidence of clinical benefit in all comers regardless of baseline biomarker levels, an unstratified Phase III trial may be pursued. The greater numbers enrolled and opportunities for additional sample collection in Phase III may enable new exploratory efforts to discover diagnostic biomarkers. However, even if these efforts are successful, a subsequent additional pivotal trial may be necessary to prospectively validate the diagnostic.

In order to have data on the biomarker in time so as not to impede the clinical drug development process, an early effort to get biomarker reagents in place is likely needed well before the start of Phase II. Even if one expects to do a pre-specified retrospective analysis of Phase II data, a commitment to the diagnostic development would need to occur in parallel with the Phase II program in order to have everything ready to be run in 'real-time' to validate the predictive effect in Phase III through prospective randomization using the commercially representative assay on the intended platform. If the drug is a new entity with no prior clinical data to mine, there is an even higher risk that one may need more than a single Phase II trial to sufficiently mitigate a risk of the test not being able to reproducibly predict a treatment benefit in the subsequent trials using the same pre-specified threshold for being 'diagnostic-positive'. Biases abound in the clinical trial setting. In addition to the inherent variability from trial to trial is the potential for other real differences in the biology between slightly different populations, such as effects from studying patients of different ages and ethnicities, with different degrees of disease severity, different co-morbid conditions, or different background medications.

Data are needed not only to understand the clinical properties of the biomarker, i.e., the specific predictive properties of the biomarker to pre-select patients most likely to benefit from the treatment. Substantial data are also required to understand the technical properties of the biomarker, i.e., the accuracy, reproducibility and real-world reliability of all aspects of the assay. Detailed information is needed on all the reagents, the equipment and laboratories intended to run the assay, as well as the analytical methods that will generate a report and how that report is to be interpreted. All of this must be done at risk early enough to troubleshoot and refine the processes, because the assay that is used for Phase III studies is expected to be the intended product, or very close to it.

These high expectations mean that a pharmaceutical company may benefit from partnering with an experienced diagnostic company. This scenario creates vulnerability for the pharmaceutical company that is not entirely within its control: if any problems occur with the test or timely access to test results, the test may become a barrier to prescribing the drug. Hence, selection of a diagnostic partner is of great importance to the pharmaceutical company. There are also considerations for competition from diagnostic labs that may try to launch competing assays. Though regulations in the US can create a high bar and should make it more difficult for a competitor to launch an equivalently labeled test, enforcement against even 'home-brew' assays in oncology indications has been fairly scarce [78,79]. Hence the mutual value proposition needs to be thoroughly assessed early in the partnership negotiation to set reasonable expectations for both sides.

Because a companion diagnostic test will be used specifically to select treatments for patients, these tests are often held to high standards for regulatory approval. In the United States, the Food and Drug Administration's Center for Devices and Radiological Health (FDA's CDRH) regulates the manufacturing, re-packaging, re-labeling and importation of medical devices, which include not only the assay reagents and platform, but also the software that analyzes the data and reports the results to treating practitioners. Devices are classified based upon the intended use and indications for use, with increasing requirements needed for an increase in the level of potential for risk posed to the patients for whom they are to be used. Unless the device meets a pre-specified exemption, Class I and II devices require acknowledgment of a '510(k)' premarket notification (21 CFR Part 807 Subpart E) before being commercially distributed. This mechanism might be acceptable for launch of a new assay that has a pre-existing assay on the market already that measures something similar. However, if a test is intended to truly be used as a companion diagnostic, to select patients for exposure to a new drug with an accompanying regulatory claim of benefit and indications for use, it will most likely need a premarket approval (PMA; 21 CFR Part 814) and will be considered a 'Class III' device because of the potential that it may pose a significant risk of harm to the patient. Furthermore if either a 510(k) or PMA will be needed, the developer may need to formally notify the agency before using the investigational device in a clinical study by submitting and seeking an investigational device exemption (IDE)(21 CFR Part 812), or by requesting a pre-submission meeting to discuss the intended use and development plan for the device. These meetings will typically involve not only the device division but also attendance by the drug division that will be reviewing the companion treatment, and can be quite informative about the expectations of the agency. For example if one intends to utilize the investigational test results to select subjects that will be included or excluded from a drug treatment study, a formal IDE may be needed to justify the design as being acceptably safe for study subjects. Regulations in different countries may be variable. In the European Union, for example, though diagnostic tests can be placed on the market in a more simplified manner by seeking a Conformité Européenne (CE) mark to self-declare that the product meets regulatory requirements, there is a mechanism for early consultation with the European Medicines Agency's Scientific Advisory Group on Diagnostics (SAG-D), which can be requested by the Committee for Medicinal Products for Human Use (CHMP) to review scientific and technical aspects of a proposed companion diagnostic program and provide advice to the companies sponsoring the combined development of the drug and device.

Although most companion diagnostic tests approved thus far have been tissue-based assays for oncology or infectious disease indications, some examples of serum protein-based diagnostics for use in inflammatory disorders like asthma are also now under development. Each platform and situation has its own challenges. As detailed in this chapter, we have worked on the development of a serum test for the measurement of periostin, a protein whose expression is linked to the presence of IL13, which has been implicated in asthma pathogenesis. Serum periostin levels were used in a pre-specified retrospective analysis of a Phase II trial to predict the clinical response of patients with asthma to lebrikizumab, an anti-IL13 biological therapy [72]. We spent many years collecting supportive data in small pilot studies without lebrikizumab to try to understand periostin biology and what a relevant threshold for a 'normal' level of periostin might be. Despite these efforts, it was recognized from the beginning that until the clinical experiment with the drug had been conducted and repeated to confirm the previous findings, the relationships between the biomarker, the drug target, the disease, and the drug would be incompletely understood at best. Though it has been challenging to provide a wealth of information about the periostin biology in health and disease in a relatively compressed timeframe to allow it to be used in an informed manner by the lebrikizumab drug development program, the lessons learned from these studies will advance the field toward better understanding a segment of asthma pathophysiology, and may redefine how we approach the treatment of patients with asthma.

4.5 OUTLOOK FOR THE FUTURE

As more clinical experience is accrued with novel asthma therapeutics and biomarkers, several key issues will need to be addressed to increase the chances of success for personalized health care in asthma. First, a clearer consensus on how asthma heterogeneity should be defined is needed, not least to foster recognition on the part of regulatory authorities that asthma is a heterogeneous disorder and that asthma heterogeneity has a molecular basis. Second, as discussed above, the promising initial findings in proof-of-concept clinical studies must be replicated and extended in larger pivotal studies. These studies will be instrumental in defining effect sizes and biomarker cutoffs more precisely. Third, as a corollary to the first two points, clinical outcome measures and biomarker assays should be standardized to the extent possible to enable better comparisons between clinical trials. Fourth, the gradients, as opposed to sharp demarcations, between asthma subtypes need to be better appreciated in the context of interventional studies. Thus far, most biomarkers used to define asthma 'subtypes' exhibit continuous rather than multimodal distributions. As such, any cutoffs applied to enable treatment decisions will necessarily require a tradeoff between the magnitude of treatment benefit and the size of the treatment-eligible population. Further complicating this issue is the possibility that different biomarker cutoffs may be optimal for different outcome measures with regard to a particular intervention. Even in the case of pharmacogenomic biomarkers defined by specific polymorphisms, it is clear that no single genetic variant will cleanly delineate asthma phenotypes; rather it is likely that many polymorphisms, each with a small effect size, contribute to the genetic component of asthma. Finally, more precise and physiologically direct outcome measures are needed to link therapeutic targets and clinical outcome measures: while FEV1 and exacerbations are valid and approvable outcomes for asthma trials, the physiological mechanisms whereby specific molecular pathways contribute to lung function in human asthma are unclear. High-resolution imaging technologies, bronchoscopies before and after treatment, and surrogate biomarkers of specific physiological processes may help to explain the molecular, cellular, and physiological mechanisms linking targets to clinical endpoints in better detail. Fortunately, the proliferation of novel agents targeting specific mediators and pathways, if rigorously examined in well designed clinical studies, will generate data to test the hypotheses underlying asthma 'endotypes', as successful therapeutic intervention in a particular pathway is necessary to confirm that pathway as a key mediator of disease.

Acknowledgments

We wish to thank Ted Rigl and Heleen Scheerens for critical review and helpful comments on the manuscript.

References

[1] Haldar P, Pavord ID, Shaw DE, Berry MA, Thomas M, Brightling CE, et al. Cluster analysis and clinical asthma phenotypes. Am J Respir Crit Care Med 2008;178:218–24.
[2] Moore WC, Meyers DA, Wenzel SE, Teague WG, Li H, Li X, et al. Identification of asthma phenotypes using cluster analysis in the Severe Asthma Research Program. Am J Respir Crit Care Med 2010;181:315–23.
[3] Anderson GP. Endotyping asthma: new insights into key pathogenic mechanisms in a complex, heterogeneous disease. Lancet 2008;372:1107–19.
[4] Sutherland ER, Goleva E, King TS, Lehman E, Stevens AD, Jackson LP, et al. Cluster analysis of obesity and asthma phenotypes. PLoS One 2012;7:e36631.
[5] Gonem S, Raj V, Wardlaw AJ, Pavord ID, Green R, Siddiqui S. Phenotyping airways disease: an A to E approach. Clin Exp Allergy 2012;42:1664–83.
[6] Hashimoto S, Bel EH. Current treatment of severe asthma. Clin Exp Allergy 2012;42:693–705.
[7] Lotvall J, Akdis CA, Bacharier LB, Bjermer L, Casale TB, Custovic A, et al. Asthma endotypes: a new approach to classification of disease entities within the asthma syndrome. J Allergy Clin Immunol 2011;127:355–60.
[8] Gibeon D, Chung KF. The investigation of severe asthma to define phenotypes. Clin Exp Allergy 2012;42:678–92.
[9] McGrath KW, Icitovic N, Boushey HA, Lazarus SC, Sutherland ER, Chinchilli VM, et al. A large subgroup of mild-to-moderate asthma is persistently noneosinophilic. Am J Respir Crit Care Med 2012;185:612–19.
[10] Simpson JL, Scott R, Boyle MJ, Gibson PG. Inflammatory subtypes in asthma: assessment and identification using induced sputum. Respirology 2006;11:54–61.
[11] Cowan DC, Cowan JO, Palmay R, Williamson A, Taylor DR. Effects of steroid therapy on inflammatory cell subtypes in asthma. Thorax 2010;65:384–90.
[12] Wenzel SE. Asthma phenotypes: the evolution from clinical to molecular approaches. Nat Med 2012;18:716–25.
[13] Woodruff PG, Modrek B, Choy DF, Jia G, Abbas AR, Ellwanger A, et al. T-helper type 2-driven inflammation defines major subphenotypes of asthma. Am J Respir Crit Care Med 2009;180:388–95.

[14] Martin RJ, Szefler SJ, King TS, Kraft M, Boushey HA, Chinchilli VM, et al. The predicting response to inhaled corticosteroid efficacy (PRICE) trial. J Allergy Clin Immunol 2007;119:73–80.

[15] Hauber HP, Gotfried M, Newman K, Danda R, Servi RJ, Christodoulopoulos P, et al. Effect of HFA-flunisolide on peripheral lung inflammation in asthma. J Allergy Clin Immunol 2003;112:58–63.

[16] Kharitonov SA, Donnelly LE, Montuschi P, Corradi M, Collins JV, Barnes PJ. Dose-dependent onset and cessation of action of inhaled budesonide on exhaled nitric oxide and symptoms in mild asthma. Thorax 2002;57:889–96.

[17] McNicholl DM, Stevenson M, McGarvey LP, Heaney LG. The utility of fractional exhaled nitric oxide suppression in the identification of nonadherence in difficult asthma. Am J Respir Crit Care Med 2012;186:1102–8.

[18] Green RH, Brightling CE, McKenna S, Hargadon B, Parker D, Bradding P, et al. Asthma exacerbations and sputum eosinophil counts: a randomised controlled trial. Lancet 2002;360:1715–21.

[19] Jayaram L, Pizzichini MM, Cook RJ, Boulet LP, Lemiere C, Pizzichini E, et al. Determining asthma treatment by monitoring sputum cell counts: effect on exacerbations. Eur Respir J 2006;27:483–94.

[20] Powell H, Murphy VE, Taylor DR, Hensley MJ, McCaffery K, Giles W, et al. Management of asthma in pregnancy guided by measurement of fraction of exhaled nitric oxide: a double-blind, randomised controlled trial. Lancet 2011;378:983–90.

[21] Braman SS. The global burden of asthma. Chest 2006;130:4S–12S.

[22] Blakey JD, Wardlaw AJ. What is severe asthma? Clin Exp Allergy 2012;42:617–24.

[23] Heaney LG, Horne R. Non-adherence in difficult asthma: time to take it seriously. Thorax 2012;67:268–70.

[24] Adcock IM, Barnes PJ. Molecular mechanisms of corticosteroid resistance. Chest 2008;134:394–401.

[25] Tantisira KG, Lasky-Su J, Harada M, Murphy A, Litonjua AA, Himes BE, et al. Genomewide association between GLCCI1 and response to glucocorticoid therapy in asthma. N Engl J Med 2011;365:1173–83.

[26] Catley MC, Coote J, Bari M, Tomlinson KL. Monoclonal antibodies for the treatment of asthma. Pharmacol Ther 2011;132:333–51.

[27] Holgate ST. Trials and tribulations in identifying new biologic treatments for asthma. Trends Immunol 2012;33:238–46.

[28] Pelaia G, Vatrella A, Maselli R. The potential of biologics for the treatment of asthma. Nat Rev Drug Discov 2012;11:958–72.

[29] Holgate ST. Stratified approaches to the treatment of asthma. Br J Clin Pharmacol 2012. doi: http://dx.doi.org/10.1111/bcp.12036.

[30] Wenzel S. Severe asthma: from characteristics to phenotypes to endotypes. Clin Exp Allergy 2012;42:650–8.

[31] Woodruff PG, Boushey HA, Dolganov GM, Barker CS, Yang YH, Donnelly S, et al. Genome-wide profiling identifies epithelial cell genes associated with asthma and with treatment response to corticosteroids. Proc Natl Acad Sci USA 2007;104:15858–15863.

[32] Holgate ST. Pathophysiology of asthma: what has our current understanding taught us about new therapeutic approaches? J Allergy Clin Immunol 2011;128:495–505.

[33] Winpenny JP, Marsey LL, Sexton DW. The CLCA gene family: putative therapeutic target for respiratory diseases. Inflamm Allergy Drug Targets 2009;8:146–60.

[34] Patel AC, Brett TJ, Holtzman MJ. The role of CLCA proteins in inflammatory airway disease. Annu Rev Physiol 2009;71:425–49.

[35] Alevy YG, Patel AC, Romero AG, Patel DA, Tucker J, Roswit WT, et al. IL-13-induced airway mucus production is attenuated by MAPK13 inhibition. J Clin Invest 2012;122:4555–68.

[36] Schroder WA, Gardner J, Le TT, Duke M, Burke ML, Jones MK, et al. SerpinB2 deficiency modulates Th1/Th2 responses after schistosome infection. Parasite Immunol 2010;32:764–8.

[37] Schroder WA, Le TT, Major L, Street S, Gardner J, Lambley E, et al. A physiological function of inflammation-associated SerpinB2 is regulation of adaptive immunity. J Immunol 2010;184:2663–70.

[38] Norris RA, Moreno-Rodriguez R, Hoffman S, Markwald RR. The many facets of the matricelluar protein periostin during cardiac development, remodeling, and pathophysiology. J Cell Commun Signal 2009;3:275–86.

[39] Takayama G, Arima K, Kanaji T, Toda S, Tanaka H, Shoji S, et al. Periostin: a novel component of subepithelial fibrosis of bronchial asthma downstream of IL-4 and IL-13 signals. J Allergy Clin Immunol 2006;118:98–104.

[40] Sidhu SS, Yuan S, Innes AL, Kerr S, Woodruff PG, Hou L, et al. Roles of epithelial cell-derived periostin in TGF-β activation, collagen production, and collagen gel elasticity in asthma. Proc Natl Acad Sci USA 2010;107(32):14170–5.

[41] Naik PK, Bozyk PD, Bentley JK, Popova AP, Birch CM, Wilke CA, et al. Periostin promotes fibrosis and predicts progression in patients with Idiopathic Pulmonary Fibrosis. Am J Physiol Lung Cell Mol Physiol 2012;303(12):L1046–56.

[42] Kudo A. Periostin in fibrillogenesis for tissue regeneration: periostin actions inside and outside the cell. Cell Mol Life Sci 2011;68:3201–7.

[43] Shikotra A, Choy DF, Ohri CM, Doran E, Butler C, Hargadon B, et al. Increased expression of immunoreactive thymic stromal lymphopoietin in patients with severe asthma. J Allergy Clin Immunol 2012;129:104–111 e9.

[44] Choy DF, Modrek B, Abbas AR, Kummerfeld S, Clark HF, Wu LC, et al. Gene expression patterns of Th2 inflammation and intercellular communication in asthmatic airways. J Immunol 2011;186:1861–9.

[45] Wenzel SE, Schwartz LB, Langmack EL, Halliday JL, Trudeau JB, Gibbs RL, et al. Evidence that severe asthma can be divided pathologically into two inflammatory subtypes with distinct physiologic and clinical characteristics. Am J Respir Crit Care Med 1999;160:1001–8.

[46] Jia G, Erickson RW, Choy DF, Mosesova S, Wu LC, Solberg OD, et al. Periostin is a systemic biomarker of eosinophilic airway inflammation in asthmatic patients. J Allergy Clin Immunol 2012;130:647–54 e10.

[47] St Ledger K, Agee SJ, Kasaian MT, Forlow SB, Durn BL, Minyard J, et al. Analytical validation of a highly sensitive microparticle-based immunoassay for the quantitation of IL-13 in human serum using the Erenna immunoassay system. J Immunol Methods 2009;350:161–70.

[48] van Wetering S, Zuyderduyn S, Ninaber DK, van Sterkenburg MA, Rabe KF, Hiemstra PS. Epithelial differentiation is a determinant in the production of eotaxin-2 and -3 by bronchial epithelial cells in response to IL-4 and IL-13. Mol Immunol 2007;44:803–11.

[49] Hoshino M, Morita S, Iwashita H, Sagiya Y, Nagi T, Nakanishi A, et al. Increased expression of the human Ca2+-activated Cl- channel 1 (CaCC1) gene in the asthmatic airway. Am J Respir Crit Care Med 2002;165:1132–6.

[50] Ben QW, Zhao Z, Ge SF, Zhou J, Yuan F, Yuan YZ. Circulating levels of periostin may help identify patients with more aggressive colorectal cancer. Int J Oncol 2009;34:821–8.

[51] Sasaki H, Dai M, Auclair D, Fukai I, Kiriyama M, Yamakawa Y, et al. Serum level of the periostin, a homologue of an insect cell adhesion molecule, as a prognostic marker in nonsmall cell lung carcinomas. Cancer 2001;92:843–8.

[52] Sasaki H, Lo KM, Chen LB, Auclair D, Nakashima Y, Moriyama S, et al. Expression of Periostin, homologous with an insect cell adhesion molecule, as a prognostic marker in non-small cell lung cancers. Jpn J Cancer Res 2001;92:869–73.

[53] Sasaki H, Yu CY, Dai M, Tam C, Loda M, Auclair D, et al. Elevated serum periostin levels in patients with bone metastases from breast but not lung cancer. Breast Cancer Res Treat 2003;77:245–52.

[54] Ignarro LJ. Nitric oxide as a unique signaling molecule in the vascular system: a historical overview. J Physiol Pharmacol 2002;53:503–14.

[55] Taylor DR, Pijnenburg MW, Smith AD, De Jongste JC. Exhaled nitric oxide measurements: clinical application and interpretation. Thorax 2006;61:817–27.

[56] Corren J. Inhibition of interleukin-5 for the treatment of eosinophilic diseases. Discov Med 2012;13:305–12.

[57] Walker W, Healey GD, Hopkin JM. RNA interference of STAT6 rapidly attenuates ongoing interleukin-13-mediated events in lung epithelial cells. Immunology 2009;127:256–66.

[58] Amerio P, Frezzolini A, Feliciani C, Verdolini R, Teofoli P, De Pita O, et al. Eotaxins and CCR3 receptor in inflammatory and allergic skin diseases: therapeutical implications. Curr Drug Targets Inflamm Allergy 2003;2:81–94.

[59] Bochner BS, Bickel CA, Taylor ML, MacGlashan Jr. DW, Gray PW, Raport CJ, et al. Macrophage-derived chemokine induces human eosinophil chemotaxis in a CC chemokine receptor 3- and CC chemokine receptor 4-independent manner. J Allergy Clin Immunol 1999;103:527–32.

[60] Hoersch S, Andrade-Navarro MA. Periostin shows increased evolutionary plasticity in its alternatively spliced region. BMC Evol Biol 2010;10:30.

[61] Flood-Page P, Swenson C, Faiferman I, Matthews J, Williams M, Brannick L, et al. A study to evaluate safety and efficacy of mepolizumab in patients with moderate persistent asthma. Am J Respir Crit Care Med 2007;176:1062–71.

[62] Juniper EF, O'Byrne PM, Guyatt GH, Ferrie PJ, King DR. Development and validation of a questionnaire to measure asthma control. Eur Respir J 1999;14:902–7.

[63] Haldar P, Brightling CE, Hargadon B, Gupta S, Monteiro W, Sousa A, et al. Mepolizumab and exacerbations of refractory eosinophilic asthma. N Engl J Med 2009;360:973–84.

[64] Nair P, Pizzichini MM, Kjarsgaard M, Inman MD, Efthimiadis A, Pizzichini E, et al. Mepolizumab for pred-nisone-dependent asthma with sputum eosinophilia. N Engl J Med 2009;360:985–93.
[65] Miller MK, Lee JH, Miller DP, Wenzel SE. Recent asthma exacerbations: a key predictor of future exacerba-tions. Respir Med 2007;101:481–9.
[66] Castro M, Mathur S, Hargreave F, Boulet LP, Xie F, Young J, et al. Reslizumab for poorly controlled, eosino-philic asthma: a randomized, placebo-controlled study. Am J Respir Crit Care Med 2011.
[67] Molfino NA, Novak R, Silverman RA, Rowe BH, Smithline H, Khan F, et al. Reduction in the number and severity of exacerbations following acute severe asthma: results of a placebo-controlled, randomized clinical trial with benralizumab. Am J Respir Crit Care Med 2012;185:A2753.
[68] Pavord ID, Korn S, Howarth P, Bleecker ER, Buhl R, Keene ON, et al. Mepolizumab for severe eosinophilic asthma (DREAM): a multicenter, double-blind, placebo-controlled trial. Lancet 2012;380:651–9.
[69] Otulana BA, Wenzel SE, Ind PW, Bowden A, Puthukkeril S, Tomkinson A, et al. A Phase IIb study of inhaled pitrakinra, an IL-4/IL-13 antagonist, successfully identified responder subpopulations of patients with uncon-trolled asthma. Am J Respir Crit Care Med 2011;183:A6179.
[70] Corren J, Busse W, Meltzer EO, Mansfield L, Bensch G, Fahrenholz J, et al. A randomized, controlled, Phase II study of AMG 317, an IL-4Ralpha antagonist, in patients with asthma. Am J Respir Crit Care Med 2010;181:788–96.
[71] Piper E, Brightling C, Niven R, Oh C, Faggioni R, Poon K, et al. A Phase II placebo-controlled study of traloki-numab in moderate-to-severe asthma. Eur Respir J 2013;41(2):330–8.
[72] Corren J, Lemanske RF, Hanania NA, Korenblat PE, Parsey MV, Arron JR, et al. Lebrikizumab treatment in adults with asthma. N Engl J Med 2011;365:1088–98.
[73] Slager RE, Otulana BA, Hawkins GA, Yen YP, Peters SP, Wenzel SE, et al. IL-4 receptor polymorphisms predict reduction in asthma exacerbations during response to an anti-IL-4 receptor alpha antagonist. J Allergy Clin Immunol 2012;130(2):516–22.
[74] Woodruff PG, Modrek B, Choy DF, Jia G, Abbas AR, Ellwanger A, et al. T-helper type 2-driven inflammation defines major subphenotypes of asthma. Am J Respir Crit Care Med 2009;180:388–95.
[75] Arron JR, Corren J, Matthews JG. Author response to correspondence on 'Lebrikizumab treatment in adults with asthma'. N Engl J Med 2011;365:2433–4.
[76] McClintock D, Corren J, Hanania NA, Mosesova S, Lal P, Arron JR, et al. Lebrikizumab, an anti-IL-13 mono-clonal antibody, reduces severe asthma exacerbations over 32 weeks in adults with inadequately controlled asthma. Am J Respir Crit Care Med 2012;185:A3959.
[77] Scheerens H, Arron JR, Choy DF, Mosesova S, Lal P, Matthews JG. Lebrikizumab treatment reduces serum periostin levels in asthma patients with elevated baseline levels of periostin. Am J Respir Crit Care Med 2012;185:A3960.
[78] FDA. Draft guidance for industry, clinical laboratories, and fda staff – in vitro diagnostic multivariate index assays. 2007. http://www.fda.gov/MedicalDevices/DeviceRegulationandGuidance/GuidanceDocuments/ucm079148.htm.
[79] Carlson B. HER2 Tests: How Do We Choose? Biotechnol Healthcare 2008;5:23–7.

microRNA Biomarkers as Potential Diagnostic Markers for Cancer

Elizabeth Mambo, Anna E. Szafranska-Schwarzbach, Gary Latham, Alex Adai, Annette Schlageter, Bernard Andruss

Asuragen, Inc., Austin, Texas

5.1 OVERVIEW OF microRNAs (miRNAs)

5.1.1 Small Non-Coding RNAs Play a Pivotal Role in Translational Control

MicroRNAs (miRNAs) are small (18–25 nucleotide) non-coding RNAs that control protein expression by binding to complementary sites on target messenger RNA (mRNA) transcripts, which results in the inhibition of the translation and/or degradation of the target mRNA [1]. Although miRNAs are present in eukaryotes, they have also been identified in organisms such as viruses and algae that are, in evolutionary distance, light years away from humans. This evolutionary conservation would suggest that miRNAs play an indispensable role in cellular functions. Indeed, they have been found to regulate critical cellular processes, such as apoptosis [2], cell proliferation [3], and cell differentiation [4,5]. More than 2000 mature, human miRNAs have been identified (http://www.mirbase.org/, release 20, July 2013) [6]. Each miRNA can regulate the expression of multiple genes, and several different miRNAs can bind to a single mRNA transcript, thereby resulting in profound effects on gene expression and cellular functions [7,8]. Due to the involvement of miRNAs in critical cellular functions, they have been implicated in a number of diseases, including cancer, in which cell proliferation and apoptosis are dysregulated. The expression of some miRNAs has been shown to be dysregulated in cancer relative to normal tissues [9,10], giving rise to the idea that miRNAs can function both as tumor suppressors and oncogenes, and are promising analytes for diagnosis, prognosis and therapy.

Y. Yao, B. Jallal, K. Ranade (Eds): Genomic Biomarkers for Pharmaceutical Development.
DOI: http://dx.doi.org/10.1016/B978-0-12-397336-8.00005-7

5.1.2 miRNAs as Potential Therapeutics

The transition of miRNAs into clinically useful therapeutics is rapidly gaining momentum. Ushering in this transition are the evolving miRNA delivery systems for overcoming the challenge of targeted delivery, and also preclinical studies which demonstrated efficacy with different miRNA-based therapies. Initial results from the first miRNA-targeted drug to enter human clinical trials are encouraging. Miravirsen (Santaris Pharma), a miRNA therapy for hepatitis C, is an oligonucleotide that targets miR-122 and uses a clever strategy to block replication of the virus. By targeting miR-122, which is highly expressed in the liver, it takes advantage of the virus' dependency on the host miRNA for viral replication. Results from a Phase IIa clinical trial with miravirsen are promising [11,12] and the findings support the potential for miRNAs therapies to have a significant impact on hepatitis infection, and potentially other infectious diseases.

miRNA-based therapies are under investigation for multiple diseases, including cancer, with potential clinical trials on the horizon. For example, miRagen Therapeutics has preclinical programs that are focused on miRNA-targeted therapeutics of cardiovascular disease and proliferative blood disorders, including chronic heart failure, post-myocardial infarction remodeling and polycythemia vera. A lead miRagen compound (MGN-4893) for treatment of polycythemia vera targets miR-451, and has the FDA's orphan drug designation. Another company, Regulus Therapeutics, reports that treatment with anti-miR-21 reduces tumor burden in preclinical models of liver cancer and results in a significant survival benefit. In 2013, Mirna Therapeutics, Inc. is expected to initiate clinical trials of a miR-34-based cancer therapy [13]. miR-34 is a well-characterized tumor suppressor that has a reduced level of expression in multiple cancer types. Mirna Therapeutics reported that a miR-34-based treatment of mice carrying existing hepatic tumors led to significant tumor regression and prolonged survival [13]. If successful, it may be one of the first miRNA mimics to reach the clinic.

5.1.3 miRNA Assays Available as Laboratory Developed Tests (LDTs)

As yet, no miRNA-based test has had FDA approval as an *in vitro* device (IVD); however, we anticipate that the field will continue to advance toward the IVD kit format. Clinical tests need to be robust, reproducible, with high sensitivity and specificity, and capable of detecting multiple miRNAs, all from clinical specimens that have a limiting amount of RNA. Rigorous testing is required for a LDT in compliance with Clinical Laboratory Improvement Amendments (CLIA) guidelines to demonstrate accuracy, precision, analytical sensitivity and specificity.

At Asuragen, we validated the first miRNA-based diagnostic test, miR*Inform*® Pancreas, as a LDT to aid in the discrimination of chronic pancreatitis from pancreatic ductal adenocarcinoma (PDAC) in formalin-fixed paraffin-embedded (FFPE) specimens [14]. Additional miRNA-based tests are in the process of clinical validation at Asuragen and we anticipate that these tests will soon be available for pancreatic cancer and other oncology applications.

Rosetta Genomics has a series of tests available (MiRview®), including a test based on the detection of 64 miRNAs that identify 42 different types of primary tumor of origin in primary and metastatic cancer. Another MiRview test differentiates squamous from non-squamous non-small cell lung cancer (NSCLC), and is able to classify squamous-cell carcinoma of the lung with 96% sensitivity and 90% specificity. In addition to a test for lung cancer which identifies its four main subtypes, another test classifies common types of kidney tumors.

5.2 DEVELOPMENT OF A miRNA-BASED CLINICAL TEST

5.2.1 Biomarker Discovery on a High Content Platform

Biomarker discovery plays a key role in the larger context of implementing a clinical test. Since the goal is to implement a validated test, the platform to be used for the test is often different from the discovery platform, and that sets the stage for a series of studies that have conceptually similar design qualities: sample size estimation; expression profiling; and performance estimation (Fig. 5.1). Sample size estimates for biomarker discovery using array technology have been extensively evaluated in the literature [15–19]. However, to our knowledge, research has not focused on incorporating the complexities and noise of assay into model migration. Migrating a predictive model from a microarray (or next-generation sequencing) platform to RT-qPCR, for example, is a challenge for clinical testing as novel design and analysis strategies for performance preservation face increased scrutiny.

A high-throughput and large-content platform is used to generate a hypothesis, typically in an unbiased manner where all possible genes (or miRNAs) are considered as possible candidates for discriminating the groups of interest. Biomarker discovery and predictive modeling has been rigorously applied to mRNA expression data derived from array technology [20–27]. Many of the qualities of sound design and analysis strategies for successful mRNA biomarker discovery also hold true for miRNA studies including:

- involvement of bioinformatics and biostatistics teams in the study design;
- balancing of classes with other relevant clinical covariates, particularly batching of experiments and site of sample origin;

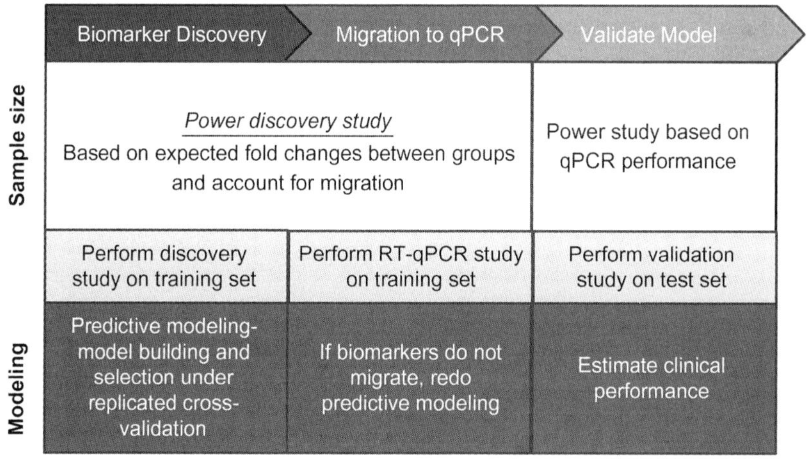

FIGURE 5.1   A high level overview of biomarker discovery in the context of platform migration. The platform used for unbiased discovery and predictive modeling is typically different than the platform used for a clinical test. As a result, a bridging study (or platform migration) occurs where results between the two platforms are shown to be equivalent. For mRNA- and miRNA-based signatures and predictive models, the destination platform is typically one based on RT-qPCR. The figure does not capture the clinical and analytical validation of the final RT-qPCR based test.

- monitoring for batch effects;
- appropriate application of replicated cross-validation including nesting of feature selection;
- application of a predictive single model to an independent test set;
- comparison of the final predictive model to a model derived from available clinical covariates [20,28–31].

The minimum prerequisites for predictive modeling include a matrix of expression values paired with (typically) dichotomized class labels, and are therefore not necessarily dependent on the origin of the expression values be it from mRNAs or miRNAs. The same could also be said for digital gene expression values derived from RNA-Seq (or microRNA-Seq) with qualifiers for regions with insufficient coverage. That is, biomarkers with insufficient coverage by RNA-Seq or low expression by microarray should be filtered from biomarker discovery efforts. Although the algorithms for classification are not as dependent on normalized expression data from different platforms, the tools for differential expression analysis are more technology specific. For example, the limma software package [32], among the most popular packages for performing differential expression analysis on microarray platforms, inspired concepts in edgeR [33], an analogous procedure for moderating mRNA (or miRNA) specific variance for digital gene expression.

Biomarker discovery is a painstakingly incremental process with numerous challenges [34]. The general aspects of bias and experimental design are critical considerations independent of the biomarker discovery platforms although greater precedent exists in the microarray literature [35]. Consistent implementation of valid methodology is essential for reproducible research, but it is often challenged by negative results [36]. Guidelines established from QUADAS and STARD can serve as useful check points along the path of discovery to clinical validity and utility [37,38].

5.2.2 Migration to a Clinical Platform

When selecting biomarkers for clinical validation, it is important to consider the changes in model performance that accompany model and signature migration to a platform with inherently different detection properties. Although the literature contains numerous examples of model stability across microarray platforms [39–43], a very common problem is migrating a model trained on microarray expression values (typically observed in the log space) to accepting inputs based on RT-qPCR data [typically observed as Ct (cycle threshold) values]. Typical solutions include mean centering of inputs across samples, but that can be complicated when samples are evaluated individually and prospectively. Another includes normalization of the microarray and RT-qPCR data in a gene-specific manner so that inputs are always as log ratios. Setting aside the platform specific interpretations of underlying analyte concentration, noise is inherently introduced by the process of model migration. This noise introduced by the process of platform migration will affect the number of samples needed for a successful biomarker discovery study and, potentially, estimates of predictive performance. Some work has focused on gene-specific behavior as opposed to model-specific behavior during platform migration [44–47], although these studies did not address how the lack of correlation between the platforms can affect the sample size estimates for biomarker

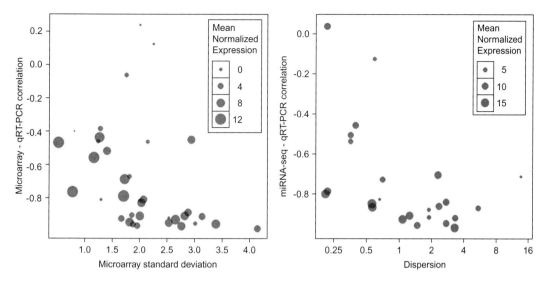

FIGURE 5.2   Association between cross-platform correlation and the standard deviation/dispersion. These plots show that expression level variation across all groups is an important factor for predicting cross-platform correlation. For both of the high content platforms (microarray, left; miRNA-Seq, right), a better correlation with PCR is accompanied by a larger total variability (within- plus between-group) in miRNA expression. This applies across the entire set and is not a within group measurement of variance. Circle size represents relative overall expression of miRNAs.

discovery or predictive model performance. In general, we expect that overall variance will drive correlation of expression signal between platforms (Fig. 5.2).

5.2.3   Validation of a Classifier

Once a model is identified and a diagnostic test is developed, the model typically undergoes a rigorous validation process to determine its analytical and clinical performance. Although validation of a model can vary depending on the clinical context, an important reference includes the guidelines published by CLIA and College of American Pathologists (CAP) for the design and development of a LDT. These guidelines prescribe the assay of specific performance parameters for the test, including sensitivity, specificity, precision, and linearity. The test validation process for the development of the miR*Inform*™ Pancreas test, which is applicable to the development of diagnostic tests in general, has been reviewed elsewhere [14], and will be briefly discussed.

Analytical sensitivity – the ability of an assay to produce a positive test result when the target is present – is evaluated by measuring the lower limit of detection and lower limit of quantification of analytes, typically titrated into a background of yeast tRNA. The analytical sensitivity is also commonly tested with RNA extracted from clinical specimens. Analytical specificity – the ability of an assay to produce a negative test result when the molecular target is not present – is evaluated by testing for non-specific amplification. For validation of the miR*Inform* Pancreas assay, analytical specificity was evaluated by comparing the amplification of synthetic miR-196a to that of closely related miR-196b, a no-template control and yeast tRNA. Targets were routinely not detected in the no-template control, while yeast tRNA targets were detected at levels that lie outside the range of reliable detection. Precision refers to the ability of an assay to produce the same result under repetition. The assessment of precision should demonstrate minimal within-run, run-to-run, lot-to-lot, and operator-to-operator variability. Finally, linearity is the ability of an assay to produce a linear relationship to known standards or dilutions. Linearity should cover the reportable range of the final test score, and the expected concentrations of target values from clinical samples. For clinical validation of a diagnostic test, the key performance parameters to be tested are clinical sensitivity and clinical specificity. Clinical or diagnostic sensitivity is the ability of an assay to produce a positive test result when the patient has the disease or clinical condition. Conversely, clinical or diagnostic specificity is the ability of an assay to produce a negative test result when the patient sample does not have the disease or clinical condition. Reasonable estimates for test results can vary from test to test, but it is critical that whatever criteria are put in place that they are well documented, reproducible and defensible.

5.2.4 miRNA Normalization

The accurate measurement of changes in miRNA expression levels is a critical objective in biomarker studies. Even modest differences in miRNA quantification can be biologically significant [48,49], and careful attention must be paid to experimental design and data interpretation to ensure reproducible measurements among samples. When comparing sample groups, such as normal and diseased biopsies, the sources of variation are manifold, and may include technical factors, such as sample procurement, handling, stabilization and storage, RNA extraction, and the efficiency of target quantification, or biological differences, such as heterogeneous and varied cellular composition. Moreover, the total miRNA population is only a fleeting fraction of the bulk RNA in a sample, and miRNA yields can vary from one specimen to the next. To address these challenges, many RNA-based assays invoke normalization as a statistical strategy to correct as much of the sample-to-sample measurement variation as possible, leaving only the biological fingerprint of miRNA expression differences to reflect phenotypic differences between experimental groups.

Because there is no such thing as a 'universal normalizer', these reference genes must be identified through empirical studies that are faithful to the experimental workflow. For RT-qPCR studies that aim to identify individual miRNA targets associated with a particular disease, a number of different approaches have been described [48–52]. Each of these methods recognizes the value of quantifying targets against endogenous RNA molecules that can correct for the noise inherent across the complex workflows that are required to reveal miRNA levels. Two of these methods deserve further discussion; these address reduced or large-scale experimental designs, respectively, and are effective for applications of miRNA biomarker discovery in diverse sample types, including challenging specimens such as biofluids.

In cases where samples are limiting, or when the highest possible performing (typically singleplex) RT-qPCR assays are desired, a reduced scale experimental plan is warranted [50]. In this approach, a short list of reference gene candidates are identified from related studies [48] or current best evidence, and these are evaluated across the samples and workflows of interest. Typically, these candidates are themselves miRNAs, since other species such as ribosomal RNA have been shown to be statistically inferior [48]. Open source programs such as geNorm [53] or NormFinder [54] are then used to identify small RNAs that represent the most stable normalizer, or more commonly, the most stable combination of normalizers. Candidate biomarkers are then assessed relative to the reference gene signal(s) to provide a reliable difference measurement between sample groups.

In contrast, experimental plans that may include highly multiplexed assays and hundreds of miRNA targets can benefit from mean centering restricted (MCR) normalization [52]. MCR normalization calculates the mean value of all detected miRNAs, and assesses expression differences against this background. An extension of this method, termed concordance correlated restricted (CCR) normalization, can be used to migrate candidates identified from MCR-corrected RT-qPCR array data to a singleplex assay format [52].

Normalization is an integral component of a larger process that must be thoughtfully constructed to achieve the goal of pinpointing miRNAs with potential as biomarkers. Since the first publication in this area in 2008 [48], multiple strategies have been devised and tested, and these methods offer flexibility across a myriad of sample and experimental conditions in pursuit of robust, disease-specific markers.

5.3 DETECTION OF miRNAs IN CLINICAL SPECIMENS FOR CANCER DIAGNOSTICS AND PROGNOSTICS

5.3.1 Pancreatic Cancer as a Model for Development of miRNA-Based Tests

To evaluate the potential of miRNAs as diagnostic tools for cancer, we assessed miRNA expression in different biopsy types including fresh frozen, FFPE, and fine needle aspiration (FNA). In this section, we will discuss the use of pancreatic cancer as a model for miRNA-based assay development. We developed the miRNA-based molecular LDT, miR*Inform* Pancreas, that can discriminate between chronic pancreatitis and pancreatic cancer to assist in obtaining the definitive diagnosis of pancreatic cancer in fixed specimens. In addition, we are validating additional assays for pancreatic cancer based on the expression of miRNAs panels for use in patients with benign, inconclusive and non-diagnostic results from endoscopic ultrasound-guided fine needle aspiration (EUS-FNA) of solid lesions, as well as an assay for optimization of surgical treatment in patients with pancreatic cystic lesions.

Despite recent significant advances in diagnostic imaging, staging, surgical techniques, and perioperative patient care, there is still a pressing need to develop diagnostic tools capable of identifying patients that are at risk, and to detect tumors in the earliest stages of cancer development to reduce the high morbidity and mortality associated with this malignancy. Only serum CA 19–9 has been recommended for use in the routine monitoring of pancreatic cancer progression in response to therapy [55,56]. However, the clinical utility of CA 19–9 is compromised by frequent increase of its levels in bile duct, gastric, and colon cancers, as well as in such non-malignant conditions as pancreatitis and cystic fibrosis, and CA-19 has only a 79–81% sensitivity and 82–90% specificity in symptomatic patients suspected of having pancreatic cancer [57].

Pancreatic cancer is the fourth leading cause of cancer-related deaths in the US, with a five year overall survival rate of 6% for all stages combined. Over half of patients are diagnosed at a late stage, when the cancer has already spread, which carries a five year survival rate of

2%. Of those who are diagnosed with local disease, there is only a 22% survival rate [58]. The most common type of pancreatic cancer, pancreatic ductal adenocarcinoma (PDAC), accounts for more than 90% of pancreatic malignancies; the less common types include the acinar cell carcinomas and endocrine tumors (1% and 5% of all pancreatic tumors, respectively).

Pancreatic cancer is often diagnosed after a mass or a dilated duct is detected in the pancreas by computed tomography (CT) or ultrasonography (US) [59]. If the mass is unresectable, EUS-FNA is typically performed to obtain a definite tissue diagnosis prior to initiation of chemotherapy or radiation. If the CT scan is indeterminate, EUS can identify smaller lesions and further define the vascular involvement [60]. An FNA is not necessary for patients who have resectable masses, as its outcome would not change the decision to proceed with the surgery. Some clinical scenarios, however, may require obtaining a confirmatory tissue diagnosis even in patients with potentially resectable disease, e.g., in patients with increased surgical risks or patients who demand a conclusive diagnosis before consenting to surgery [61]. The sensitivity and specificity of these imaging modalities range as follows: dual phase helical CT: 78.6–98% and 54–100%, respectively; transabdominal US: 77.9–83% and 99–100%, respectively; EUS-FNA: 79.4–92% and 97–100%, respectively [62–68].

The EUS-FNA procedure has emerged as a very specific and minimally invasive modality in the pre-operative diagnosis and staging of pancreatic cancer. However, approximately 30% of EUS-FNA specimens from the pancreas are estimated to be unsuitable for diagnostic use [69–75]. In these cases, a biomarker that can distinguish pancreatic cancer cells from reactive ductal epithelium would be particularly beneficial to reduce the number of FNA passes required to obtain an affirmative diagnosis. The EUS-FNA sampling method yields material of sufficient quality and quantity for miRNA biomarker discovery studies [76].

5.3.1.1 *Improving the Diagnosis of PDAC*

One of the major clinical dilemmas is distinguishing pancreatic cancer from benign conditions, such as chronic pancreatitis, as both diseases can present with similar clinical symptoms and imaging features. In addition, the atypical cytological changes associated with pancreatitis, such as irregular separation and distortion of the actively growing connective tissue in existing pancreatic ducts, may mimic some changes seen in cancer. It has been reported that up to 25% of the cases suspected to be malignant from cytology were found to be benign at surgery [77–79]. Routine imaging techniques alone, such as CT or MRI, can neither detect PDAC at early stages nor differentiate between a benign reactive gland of chronic pancreatitis and an infiltrating gland of well-differentiated pancreatic cancer, especially in patients with underlying chronic pancreatitis [80].

Because PDAC may develop over 15–20 years, the prospect of routine testing and follow-up is compelling [81]. Molecular tests with enhanced sensitivity will improve the diagnostic accuracy of EUS-FNA, as well as improve survival rate by enabling the detection of the disease at earlier stages [82]. An important step toward achieving that goal is increasing our understanding of the biology of pancreatic carcinoma through miRNA expression studies. As an example, miR-155 emerged as a repressor of tumor protein 53 (p53)-induced nuclear protein 1 (TP53INP1) expression, leading to enhanced tumor-forming capacity of cells, while miR-34a was shown to be directly trans-activated by p53 and to promote apoptosis [83,84]. With the mounting evidence that miRNAs regulate key components of pathways associated with development of pancreatic cancer, it is not too surprising that they may also become important tools for diagnosis and classification of this disease. In fact, a number of early studies identified distinct miRNA expression profiles of pancreatic cancer and non-malignant pancreatic diseases using frozen and FFPE tissues, leading to identification of key potential diagnostic miRNAs.

A small number of candidate miRNAs, including miR-196a, miR-196b, miR-135b, miR-150, miR-155, miR-21, miR-210, miR-221 and miR-222, have been reported as significantly elevated in PDAC, with a much lesser contribution from those that are down-regulated (e.g., miR-130b, miR-148a, miR-148b, miR-216, miR-217, and miR-96) [85–91]. Detection of some miRNAs has been associated with different stages of pancreatic intraepithelial neoplasias (PanINs), the precursor lesions involved in the progression to PDAC. For example, expression of miR-196a was shown to progressively increase with the PanIN grade, with peak production occurring at the stage of PanIN-2 and -3 [92]. The closely related miR-196b was shown to be specifically over-expressed in PanIN-3 lesions (carcinoma *in situ*) as compared to lower-grade PanINs and normal pancreas [93], while miR-148a was reported to be down-regulated earlier in the course of pancreatic carcinogenesis (PanIN-1b) via hypermethylation of the DNA region encoding miR-148a, and persists at that level through the lesion's progression into PDAC [94].

Notably, miR-196a along with miR-217, a marker uniquely associated with normal pancreatic cells, became crucial components of the first miRNA-based LDT, the miR*Inform* Pancreas test. This test was developed using FFPE specimens to measure the increased proportion of ductal adenocarcinoma cells (indicated by up-regulation of miR-196a) relative to the decline in the number of acinar cells observed in PDAC (resulting in reduced expression levels of miR-217). The score was defined as the difference in expression (ΔCt) between miR-196a and miR-217. The specimens were classified as either benign or PDAC according to a cutoff point of 0.5 ΔCt, where $\Delta Ct > 0.5$ indicated a diagnostic negative (benign) and $\Delta Ct \leq 0.5$ indicated a diagnostic positive (PDAC). During the clinical validation that was performed in accordance with CLIA and CAP guidelines, the miR*Inform* Pancreas test showed a high sensitivity and specificity of 95% (95% CI: 76–100) and 95% (95% CI: 83–99), respectively, for diagnosis of PDAC [14]. Although this test has a limited clinical utility because it is based on resected tissue, it demonstrates the diagnostic capability of miRNAs confirmed by the gold standard: histology on resected tissue.

5.3.1.2 *Improving Diagnosis and Treatment of Cystic Lesions*

Pancreatic cystic neoplasms are discovered more frequently because of the increase in the use of high-resolution radiological imaging for the evaluation of abdominal trauma and other disorders [95]. The prevalence of unsuspected, asymptomatic pancreatic cysts identified during imaging can be as high as 2.6% [96]. Intraductal papillary mucinous neoplasms (IPMN) are the most common cystic precursor lesions of PDAC. However, current pre-operative modalities including imaging, endoscopy, and biochemical cyst fluid analysis fail to accurately assess the cysts' grade of dysplasia and presence of invasion without resorting to resection and precise histopathologic examination. For example, CT and MRI imaging features have only a 50% diagnostic accuracy. EUS can only differentiate mucinous from non-mucinous lesions in about 50% of the cases [97–100], which is important because mucin is characteristic of lesions that have malignant potential, such as IPMN and mucinous cystic neoplasms (MCN). Cytologic evaluation of aspirated cystic fluid from FNA is often performed during EUS, but it has at best 50% overall accuracy, which decreases dramatically in smaller pancreatic cysts [98,101–103].

Detection of glycoprotein tumor antigens, such as carcinoembryonic antigen (CEA), secreted by the epithelium lining of mucinous lesions, has modest sensitivity and specificity of 75% and 84%, respectively, for mucinous cystic lesions with CEA >192 ng/ml. However, many mucinous lesions with CEA <192 ng/ml are missed using this cutoff [98,104]. In addition, the CEA level in cyst fluid is not predictive of malignancy [105]. Furthermore, recent analyses of DNA mutations in cyst fluid have similarly proved to be disappointing. Measurement of allelic loss amplitude showed a sensitivity of 67% and a specificity of 66% for mucinous cystic lesions. The presence of KRAS mutations, although highly specific (96%) for mucinous lesions, showed a low sensitivity of 45% [106].

This persistent diagnostic uncertainty leads to treatment decisions that result in unnecessary life-altering surgeries in patients who have cysts with low grade (LG) dysplasia and who could be managed by observation. Consequently, more accurate diagnostic tools are needed to improve the treatment of patients with pancreatic cysts. In particular, the ability to preoperatively predict the grade of dysplasia is poised to have a direct impact on subsequent management, since IPMN with LG dysplasia can potentially be managed conservatively, while those with high grade (HG) dysplasia would require surgical intervention due to the high probability of invasive neoplasia.

A recent study by Wu et al. applied massively parallel DNA sequencing to evaluate the mutation status of 169 genes commonly altered in human cancers in cystic fluid specimens collected immediately after resection, and found mutations in the GNAS gene at codon 201 in approximately two thirds of IPMN, but not in serous cystadenoma (SCA) or MCN [107]. The combination of GNAS and KRAS mutations allowed the researchers to distinguish between SCA and IPMN with 96% sensitivity (95% CI, 0.91 to 0.99) and 100% specificity (97.5% one-sided CI, 0.92 to 1). The presence of a GNAS mutation in cyst fluid could also discern IPMN from MCN, albeit with a lower accuracy due to the presence of KRAS mutations in approximately 30% of MCN specimens. Although diagnostic for IPMN, the GNAS/KRAS biomarker combination does not allow for the prediction of the grade of dysplasia or presence of malignancy. Thus, there is still an urgent need for identifying cyst fluid biomarkers that would be predictive of HG dysplasia or invasion, thus enabling the stratification of patients for surgery.

The solution to this clinical dilemma may be provided by the application of miRNAs. The feasibility of using miRNAs in pancreatic biofluids was evaluated by many research groups investigating candidates previously shown to be associated with development of pancreatic cancer. For example, expression of miR-155 was interrogated in pancreatic juice samples and found to be elevated in 60% of IPMN specimens as compared to none of the disease controls [108]. In pancreatic cystic fluid, miR-21, miR-221, and miR-17–3p were able to differentiate the mucinous cysts from those nonmucinous specimens with p<0.01 [109]. The key candidate, miR-21, was able to resolve those diagnostic entities with a median specificity of 76% and a sensitivity of 80%. Recently, Matthaei et al. used cyst fluid specimens collected after surgical resection to build a miRNA classifier, which allows prediction of grade of dysplasia within the lining epithelium of an IPMN and identifies cysts that likely need surgical removal [110]. They comprehensively evaluated expression levels of up to 750 miRNAs in parallel studies, involving microdissected LG and HG FFPE IPMN specimens and cystic fluid specimens representing these diagnostic entities. They identified a subset of 18 miRNAs that segregated cysts typically requiring surgery (e.g., HG IPMN) from less malignant lesions that can be conservatively managed (e.g., LG IPMN and SCA). Subsequently, they were able to develop a logistic regression model comprised of nine miRNAs (miR-18a, -24, -30a-3p, -92a, -342-3p, -99b, -106b, -42-3p, and -532-3p), which predicted with 89% sensitivity and 100% specificity which cyst fluid specimen came from a cyst requiring resection or more conservative management.

The studies suggest that miRNA signatures may complement KRAS/GNAS combination and become a powerful tool that can help clinicians achieve a confirmatory diagnosis and guide patient management decisions (Fig. 5.3). However, prospective multi-center testing of signatures using EUS-FNA specimens will be essential to confirm their clinical application.

5.3.1.3 Prognosis

Prognosis for patients with pancreatic cancer can be affected by many factors, such as the cancer type, stage and grade (indicative of aggressiveness and metastases), the patient's age and general health, or the effectiveness and risk of treatment options. Studies interrogating prognostic capabilities of miRNAs have yielded a handful of candidates that predict patient outcome.

Perhaps the most studied prognostic miRNA candidate is miR-21. Patients with pancreatic cancer and high expression of miR-21, as compared to patients with low expression of miR-21, have a shorter survival in both the metastatic and the adjuvant settings, and are often resistant to gemcitabine treatment [88], [111]. Furthermore, miR-21 expression levels in PDAC tissue specimens were shown to be correlated with pre-operative serum levels of CA 19-9 ($r = 0.63$, $p = 0.0022$) [112]. Similarly, increased expression levels of miR-196a-2 in tissue [86] and serum [113] specimens from patients with PDAC were associated with a significantly shorter median survival as compared to patients with low levels of those miRNAs, and predicted patient tumor resectability [113]. Another study reported that low levels of miR-10b in FNA samples collected from patients who underwent neoadjuvant therapy were associated with improved response to gemcitabine-based neoadjuvant therapy, longer time to metastasis and improved overall survival of 50% after two years [114]. It appears that miR-10b could be used as an indicator of localized pancreatic disease, similarly to miR-143 expression which was negatively correlated with lymph nodes spreading ($r = -0.64$; $p = 0.0004$) [112]. Two groups reported miRNA signatures that showed utility in predicting survival in patients with nodal disease. The miRNAs in these signatures did not overlap and included: miR-452, miR-105, miR-127, miR-518a-2, miR-187 and miR-30a-3p [86], as well as miR-155, miR-203, miR-210 and miR-222 [91].

5.3.2 miRNAs in Biofluids

Cancer is currently the third leading cause of human mortality after heart disease and infectious diseases. However, cancer mortality is expected to surpass that of cardiovascular diseases by the year 2020 [115]. Evidence has shown that most cancers including breast, colorectal (CRC) and melanoma are treatable if detected early. Consequently, the development of non-invasive, early-detection biomarkers for the diagnosis of multiple human cancers has been a major area of focus.

Circulating biomarkers which are present in blood offer advantages over image-based diagnostics, such as the following:

1. They are minimally invasive.
2. They can be monitored frequently over time in a subject to establish an accurate baseline.
3. They have relatively low cost compared to imaging procedures.

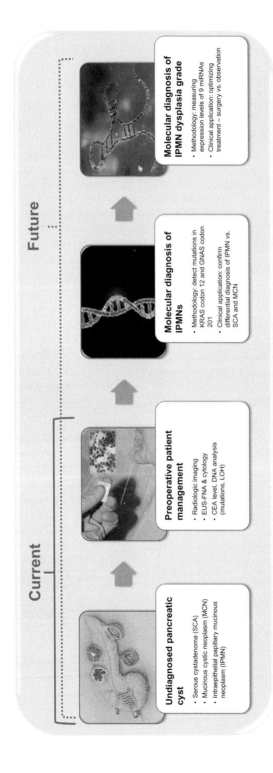

Undiagnosed pancreatic cyst

· Serous cystadenoma (SCA)
· Mucinous cystic neoplasm (MCN)
· Intraepithelial papillary mucinous neoplasm (IPMN)

Preoperative patient management

· Radiologic imaging
· EUS-FNA & cytology
· CEA level, DNA analysis (mutations, LOH)

Molecular diagnosis of IPMNs

· Methodology: detect mutations in KRAS codon 12 and GNAS codon 201
· Clinical application: confirm differential diagnosis of IPMN vs. SCA and MCN

Molecular diagnosis of IPMN dysplasia grade

· Methodology: measuring expression levels of 9 miRNAs
· Clinical application: optimizing treatment – surgery vs. observation

Current   Future

FIGURE 5.3   Current and proposed future clinical management of patients with pancreatic cystic lesions.

4. They eliminate the radiation exposure risk associated with the use of CT scans.
5. Unlike CT scans, they may allow the distinction of indolent from more aggressive lesions.

In addition, circulating biofluids offer yet another advantage of enabling cost-effective monitoring.

In 2008, the detection of circulating miRNAs in serum from diffuse large B-cell lymphoma [116] provided a new class of potential biomarkers for cancer and other diseases. In the same year, Mitchell et al. [117] showed that miRNAs were present in plasma, and that serum miRNAs could enable the distinguishing of prostate cancer patients from healthy controls. Since then, several studies have shown that circulating miRNAs have great potential to serve as diagnostic markers for various cancer types including breast, colon, ovarian, pancreatic and lung cancer (reviewed in [118]).

In general, circulating nucleic acids are believed to exist in two types of cell-derived lipid vesicles: micro-vesicles of ~100 nm – 1 μm in diameter, and exosomes of ~30–100 nm [119–122]. Because blood is enriched with enzymes that degrade RNA [123], it is suggested that the vesicles/exosomes protect circulating miRNA from degradation [124,125]. Evidence indicates that circulating miRNAs could function in cell-to-cell communication [126]; however, the specific functions and mechanisms of circulating miRNAs remain largely elusive. In spite of that, circulating miRNAs have recently emerged as potential biomarkers for cancer diagnostics and prognostics (for review see [127]). In fact, miRNAs have been further proposed as tools for cancer monitoring and treatment of several cancer types, through miRNA replacement similar to gene therapy. In this chapter, we discuss the potential utility of miRNAs from blood, serum, plasma, and urine as diagnostic biomarkers in several human cancer types. We also discuss the challenges that must be overcome in order to bring circulating miRNAs into clinical practice. This discussion will focus on cancers in which circulating miRNAs have been implicated. Circulating miRNAs have also been reported in other cancers such as renal, ovarian and head and neck cancer; however, these cancers have not been reviewed in this chapter due to the limited number of studies which have been carried out.

5.3.2.1 Leukemia

The ability of miRNA to distinguish cancer from normal tissues was first demonstrated in chronic lymphocytic leukemia (CLL), in which miR-15a and -16 were both shown to be down-regulated in CLL patients relative to cancer-free subjects, indicating that both miRNAs may act as tumor suppressors [128]. The presence of both miR-15 and -16 on chromosome 13q14, which is deleted in the majority of CLL cases, further supported the role of the two miRNAs. In 2004, Calin et al. showed that several miRNAs including miR-10b, 181b and others were de-regulated in CLL [129]. Further evidence showed that reduced expression of miR-29b and miR-181b was associated with poor prognosis in CLL [125], suggesting that like miR-15b and miR-16, miR-29b and miR-181b also act as tumor suppressors. In 2006, Pekarsky et al. showed that oncogene T cell leukemia/lymphoma 1 (TCL1) is regulated by miR-29b, and that TCL1 expression was inversely proportional to that of miR-29b, and further that high TCL1 levels were correlated with high ZAP-70 [130].

The discovery that miRNAs were differentially expressed in CLL set the stage for the study of miRNAs in other forms of leukemia, including acute lymphoblastic leukemia (ALL) and acute myeloid leukemia (AML). Similarly to CLL, miR-15b and miR-16 were found to be important miRNAs in the serum of ALL patients, along with miR-24 and miR-21 [116,117]. Tanaka et al. showed that differential expression between miR-92a and miR-638 in plasma could distinguish ALL patients from healthy control subjects, suggesting that the miRNA pair could have clinical applications in ALL [131].

There are a limited number of studies of miRNAs in AML. Costinean et al. showed that miR-155 was up-regulated in AML [132]. Mi et al. compared miRNA expression profiles in ALL and AML and found several differentially expressed miRNAs between the two, with a number of miRNAs found to be expressed at a much higher level in ALL than AML including miR-128a, miR-128b, miR-151*, miR-5, miR-130b, and miR-210 [133]. These findings indicate that not only can miRNAs distinguish cancer from normal samples, but that they are also expressed in a disease-specific manner. Schotte et al. corroborated Mi's finding showing that miRNA expression profiles in leukemia were expressed in a subtype-specific manner [134]. Because leukemia is a blood disorder, it is reasonable to envision circulating miRNAs as biomarkers for the disease. In contrast, less is known about circulating miRNAs in patients with solid tumors and their potential as diagnostic biomarkers. However, miRNA research in leukemia has paved the way for the study of circulating miRNA in solid tumors, and the findings of some of these studies are discussed below.

5.3.2.2 *Pancreatic Cancer*

Blood biomarkers may provide a minimally invasive and cost-effective way of detecting pancreatic cancer; however, as discussed in the previous section, there is currently no marker with sufficient diagnostic sensitivity and specificity to identify pancreatic cancer patients early. Several research groups assessed the feasibility of using plasma for the detection of miRNAs that were previously reported to be aberrantly expressed in PDAC, including miR-21, miR-210, miR-155, and miR-196. Wang et al. reported that the combination of these four miRNAs detected in plasma could discriminate PDAC from healthy individuals with a sensitivity of 64% and specificity of 89% [135]. Another group evaluated the supplementary role of miRNAs in plasma to serum measurements of CA 19–9 in early identification of patients with PDAC [136]. Out of the seven interrogated miRNAs (miR-16, miR-21, miR-155, miR-181a, miR-181b, miR-196a, and miR-210), miR-16 and miR-196a independently distinguished between PDAC and chronic pancreatitis, and in combination with CA 19-9 discriminated patients with PDAC from a combination of normal controls and chronic pancreatitis with a sensitivity of 92.0% and specificity of 95.6%, and from chronic pancreatitis only with sensitivity of 88.4% and specificity of 96.3%.

Analysis of miRNA expression levels in serum collected from patients diagnosed with PDAC led to development of another miRNA signature consisting of miR-20a, miR-21, miR-24, miR-25, miR-99a, miR-185, and miR-191 [137]. In the study validation phase, the diagnostic sensitivity and specificity of this 7-miRNA classifier was 94% and 93%, respectively. With the exception of miR-21, there were no other miRNAs that overlapped with those identified in human plasma, emphasizing the need for molecular diagnostic tools to be developed on a specimen that is relevant to the clinical question posed.

Finally, a study exploring the potential value of seven fecal miRNAs (miR-16, miR-21, miR-155, miR-181a, miR-181b, miR-196a, and miR-210) in screening for pancreatic cancer was reported [138]. miR-181b and miR-210 discriminated PDAC from normal individuals with receiver operating characteristic (ROC) curves and area under the curve (AUC) of 0.745 and

0.772, respectively. In addition, a significant correlation between miR-196a and the maximum tumor diameter (Spearman r = 0.516, p = 0.041) was uncovered.

These exciting 'proof of principle' studies show the feasibility of developing sensitive and specific blood-based miRNA molecular testing for detection of pancreatic cancer. These tools could eventually be applied clinically, provided that their performance characteristics can be confirmed in independent, well-controlled, large prospective multi-center studies.

5.3.2.3 Lung Cancer

The majority (70%) of all lung cancer diagnoses are detected at late stages, involving regional or distant metastatic disease. This late diagnosis contributes to a high mortality rate, with an overall five year survival rate of less than 15% [139]. Unlike CRC and breast cancer, which have screening programs [140,141], lung cancer does not have an approved annual screening method. A recent prospective study by the National Lung Cancer Screening Trial (NLST) revealed a 20.3% decrease in lung cancer mortality in the low-dose CT screening arm of the trial compared to the control arm screened by chest X-ray [142]. As a result, the National Comprehensive Cancer Network has updated its screening guidelines to include low-dose CT for those at risk of lung cancer. However, there are challenges in terms of over-diagnosis due to the detection of otherwise non-lethal/benign lesions of the lung. Apart from the high cost, there are concerns about risk of repeated exposure to radiation, as positive CT results have to be followed by periodic CT scans every six months or more often if needed. Therefore, there is an urgent need to identify biomarkers that can be used as alternative non-invasive screening tools. Differential expression of miRNAs in lung cancer tissues compared to normal lung tissue has been extensively studied and is well-demonstrated, especially in the case of the let-7 family of miRNAs [143]. However, only in recent years have circulating miRNAs from biofluids, such as whole blood, plasma, serum, and sputum, been examined as potential diagnostic markers for lung cancer [144,145]. Some noteworthy studies on circulating miRNAs in lung cancer are highlighted below.

A study by Shen and coworkers involving a total of 86 NSCLC and 57 controls revealed a four-miRNA panel in plasma, including miR-126, which distinguished NSCLC from healthy controls with a sensitivity and specificity of 73% and 96%, respectively [146]. Using whole blood, Keller and coworkers showed that miR-126 and miR-98 were among the top miRNAs that could distinguish NSCLC from healthy controls [147]. Other recent studies by Foss and coworkers showed serum miR-1254 and miR-574-5p were differentially expressed between NSCLC (N = 33) and healthy controls (N = 42) [148]. Work by Chen et al. showed that miR-205, miR-25, and miR-223 were up-regulated in serum from NSCLC patients when compared to healthy controls [149]. In 2009, Rabinowits and coworkers reported elevated levels of miR-155 and miR-21 in circulating tumor-derived exosomes in the plasma of lung cancer patients [150]. Using 50 serum specimens (20 healthy controls and 30 NSCLC) as the training set and 130 specimens (75 healthy controls and 55 NSCLC), we showed that the differential expression of miR-15b and miR-27b was able to discriminate NSCLC from healthy controls [151]. Using a cross-validation method, Patnaik et al. showed that miR-190b, miR-630, miR-942, and miR-1284 in whole blood were the most commonly used miRNAs in the classifier generated for distinction of lung adenocarcinoma from healthy controls [152].

In yet another recent study, Wang and coworkers revealed altered expression of five miRNAs (miR-93, miR-100, miR-151, miR-134, and miR-345) in pleural effusions of 184 NSCLC patients, and further that these miRNAs were associated with poor survival [153]. On the other hand, Yuxia et al. examined serum levels of miR-125b in 193 NSCLC patients before and after surgery and therapy, and found that serum miR-125b distinguished NSCLC patients from the control group with an AUC of 0.786 [154]. In addition, multivariate analysis revealed that high expression of miR-125b was significantly ($p < 0.0001$) associated with poor prognosis. Le et al. also examined serum miRNAs from pre- and 10 days post-surgery in 82 NSCLC patients, and found that levels of miR-21, miR-205, miR-30d, and miR-24 were increased in those with lung cancer relative to controls [155]. Additionally, levels of miR-21 and miR-24 were reduced in post-operative samples compared to pre-operative serum samples.

miRNAs have not only been shown to be useful as diagnostic markers, but miRNA expression levels have also been demonstrated to change with cancer progression. For example, Zheng et al. showed that miR-155 and miR-197 were higher in the plasma of lung cancer patients with metastatic disease than those with localized tumors [156]. As discussed above, a successful biomarker in lung cancer must have the ability to distinguish benign lung tumors from invasive tumors. miRNAs have also shown promise in this area, with Shen et al. having showed that a combination of three plasma miRNAs (miR-21, miR-210, and miR-486–5p) could distinguish malignant solitary pulmonary nodules from benign nodules [157]. Roth and coworkers showed that serum levels of miR-10b, miR-141 and miR-155 could also distinguish lung cancer from benign disease [158]. Similarly, Xie et al. showed that serum miR-24, miR-26a, and miR-30d could distinguish malignant form benign lung disease, in addition to predicting response to docetaxel [159].

In this era of emerging personalized medicine, there has been an increased interest in predictive biomarkers. miRNAs have been shown to predict response to chemotherapy in NSCLC patients. Specifically, Wei et al. examined miR-21 in 30 healthy controls and 63 lung cancer patients, including 11 with partial chemotherapy response and 24 with stable and progressive disease [160]. miR-21 was not only found to be diagnostic, but also highly associated with response to platinum-based chemotherapy response. The group with partial response had much lower levels of miR-21 than the group that included patients with stable disease and patients with progressive disease. In a study involving 78 NSCLC patients and 48 controls, Silva et al. revealed five plasma miRNAs (let-7f, miR-20b, miR-30e-3p, miR-223, and miR-301), combinations of which were either diagnostic or prognostic for NSCLC [161]. While all of the studies on circulating miRNA in lung cancer highlight the great promise and potential of miRNAs as diagnostic, prognostic, and even predictive biomarkers, the lack of consensus among the results is worrisome and is discussed in the final section of this review regarding the challenges of miRNAs as biomarkers.

5.3.2.4 *Breast Cancer*

Breast cancer, the leading cause of cancer mortality in women worldwide [162], has been one of the most studied cancers in terms of circulating miRNA. Although breast cancer mortality has been significantly reduced with the introduction of mammograms [163], about 30% of the cases are still diagnosed at a stage that involves regional lymph nodes, with ~62% being detected at a localized stage [164]. In addition, in some parts of the world, women have limited access to annual mammograms. Consequently, the development of a cost-effective, early-detection assay for breast cancer is still an area of active research.

CEA is a circulating tumor marker that is used as a prognostic marker as well as a mechanism to monitor recurrence. However, the poor sensitivity of CEA makes it less than ideal [165]. Several investigators have examined the potential role of circulating miRNAs as diagnostic markers for breast cancer, and most of this work has emerged in the last two years. Changes in circulating miRNA expression levels have been associated with tumorigenesis, prognosis, and response to therapy.

Schrauder and coworkers evaluated the expression of 240 miRNAs in whole blood of 48 early-stage breast cancer patients and 57 healthy controls, and reported that miR-202 was differentially expressed between the two groups [166]. Zhao et al. examined plasma miRNAs in 93 breast cancer patients who previously received neoadjuvant chemotherapy, and 32 healthy volunteers [167]. Using a cohort of serum from 102 pre-operative breast cancer patients, 34 post-operative samples, 32 with benign breast disease, and 53 healthy controls, Schwarzenbach and coworkers showed that miR-20a and miR-21 were expressed at higher levels in women with breast cancer and benign disease compared to healthy women [168]. However, only miR-214 was able to distinguish breast cancer from benign disease and controls, with an AUC of 0.878 and 0.883, respectively. Furthermore, miR-214 levels were reduced in post-operative sera. miR-214 has been shown to target PTEN tumor suppressor gene [169]. Van Schooneveld et al. selected the most differentially expressed miRNAs after examination of 84 breast cancer and eight normal breast tissues, and analyzed serum from 75 breast cancer patients and 20 healthy controls [170]. The study revealed that miR-215, miR-411, and miR-299 were differentially expressed between sera from healthy controls and that from patients with untreated metastatic breast cancer.

The ability of miRNAs to predict therapy response has been recently evaluated in breast cancer. Specifically, Jung et al. evaluated plasma miRNA expression in breast cancer patients who received neoadjuvant trastuzumab-based chemotherapy including 18 with complete response and 11 with residual disease, in addition to 39 pre-operative and 30 post-operative sera from breast cancer patients who did not receive neoadjuvant therapy [171]. Their findings revealed that miR-210 was expressed at higher levels in patients with residual disease than those with a complete response. In addition, miR-210 was higher in pre- and post-surgery in patients with lymph node involvement, suggesting that miR-210 could be used as a biomarker to predict response to trastuzumab-based therapy. In another study, Zhao et al. found that miR-221 was significantly associated with hormone receptor status, with high miR-221 expression being associated with negative hormone receptor status [167]. These studies suggested that miR-221 could be a useful predictive biomarker for neoadjuvant chemotherapy. miR-122 and miR-375 were also revealed as predictive markers of neoadjuvant chemotherapy in a study by Wu et al. that examined pre-treatment sera from 42 breast cancer patients with stage II and III disease, and validated the findings in an independent set of 26 sera samples [172].

In yet another predictive miRNA study, Wang and coworkers showed that high expression of miR-125b in serum of 56 breast cancer patients treated with neoadjuvant therapy was associated with poor response in 26 patients [173]. This high expression of miR-125b was also associated with a lower apoptotic rate and a higher percentage of proliferating cells. One of the largest studies in circulating miRNAs in breast cancer is that of Cuk et al., who examined plasma from 80 healthy controls and 127 sporadic breast cancer cases, and found that miR-148b, miR-376c, miR-409-3p, and miR-801 were significantly up-regulated in plasma from breast cancer patients, suggesting that these miRNAs could serve as useful screening markers for breast cancer [174].

More recently, another large study involving plasma from over 269 breast cancer patients [with and without circulating tumor cells (CTC)], including 76 controls, showed that miR-200b could distinguish plasma of circulating tumor cell (CTC)-positive patients from that of CTC-negative patients [175]. The association of miR-200b with CTC, an established prognostic marker for metastatic breast cancer, suggests that this miRNA could be also be used as a prognostic marker for breast cancer. Other miRNAs shown to be reportedly higher in the CTC positive patients include miR-141, miR-20a, miR-200c, miR-375, miR-203, miR-210, and miR-801. miR-21 is one of the most commonly up-regulated miRNA in most cancer types, and it was expressed at higher levels in sera of breast cancer patients than in sera of controls [176]. Similarly to the different sets of circulating miRNAs that have been identified in lung cancer patients, it is notable that there is no consensus in the circulating miRNA findings in breast cancer.

5.3.2.5 Gastric Cancer

Gastric cancer, commonly referred to as stomach cancer, is a highly prevalent type of cancer in most parts of the world. It is one of the most common causes of cancer death worldwide and causes about 738,000 deaths annually [177]. Part of this high mortality rate may be due to the fact that the molecular mechanisms leading to gastric cancer are not yet fully understood. miRNAs have been implicated in gastric cancer and associated with cancer type, stage, and survival [178]. Ueda et al. identified differentially expressed miRNAs in 160 paired tumor and normal tissues from gastric patients [179]. An analysis by Valladares-Ayerbes et al. of the dataset published by Ueda et al. revealed that miR-200 was not differentially expressed in the paired non-tumor and cancer samples; however, they found that miR-200c was significantly (p = 0.012) higher in blood samples from gastric cancer patients (N = 52) compared to controls (N = 15) [180]. The increase in miR-200c levels was also associated with poor overall survival, suggesting that miR-200c is a potential circulating prognostic marker for gastric cancer. Another potential diagnostic and prognostic miRNA marker, miR-196a, was reported by Tsai et al [181]. miR-196a was found to be up-regulated in gastric cancer tissues compared to normal tissues, and high levels of serum miR-196a was associated with recurrence of gastric cancer [181].

Circulating miRNA markers with prognostic and therapeutic potential for gastric cancer have also been identified. Wang et al. identified miR-17-5p and miR-20a in pre- and post-operative gastric cancer patients, and reported that the miRNA pair was associated with differentiation status and TNM stage of gastric cancer [182]. Wang et al. also showed that, in a mouse tumor model, tumors regressed after the mice were treated with antagomirs (synthetic oligonucleotides designed to reduce or abolish miRNA activity) specifically targeted at miR-17-5p and miR-20a. Serum levels of the two miRNAs were also reduced in mice that were treated and that showed tumor regression [182], further suggesting that this pair of miRNAs could also be used for monitoring therapy response. In an effort to identify diagnostic markers for gastric cancer, Liu et al. used qRT-PCR assays to screen serum of gastric cancer patients and controls, and reported that miR-378 was able to distinguish gastric cancer from controls with an AUC of 0.86 and sensitivity and specificity of 87.5% and 70.73%, respectively [183]. miR-21 has been shown to be up-regulated in most cancers, and exerts its oncogenic activity by targeting a number of cancer suppressor genes [184–186]. miR-21 was reported to be elevated in peripheral blood of gastric cancer patients compared to controls, and it was further reported that this miRNA was associated with TNM stage [187].

Zhou et al. spiked gastric cancer cells into blood samples from healthy volunteers and showed a correlation between the number of gastric cancer cells and the levels miR-106a and miR-17 [188]. These two miRNAs were significantly ($p<0.05$) higher in peripheral blood of pre- and post-operative gastric patients (N = 90) relative to controls (N = 27). Studies by Tsujiura and coworkers corroborated the miR-106a findings of Zhou et al. [188] in addition to reporting other miRNAs like miR-106b, miR-21, and miR-17-5p as being significantly up-regulated in plasma of gastric cancer patients (N = 69) relative to controls (N = 30) [189]. Interestingly, Tsujiura and et al. also found that the expression levels of let-7a were lower in gastric cancer patients relative to controls ($p = 0.002$). The miRNA pair, miR-106a and let-7a, was able to distinguish gastric cancer patients from controls with an AUC of 0.879, which was better than that of miR-106a alone (AUC = 0.721). However, with the exception of miR-21, miR-106a, and miR-17-5p, there is very little consensus on the potential diagnostic circulating miRNAs in gastric cancer.

5.3.2.6 Colorectal Cancer

Colorectal cancer (CRC) is one of the major cancer types that affects both men and women. In the US, there are over 143,000 new cases of CRC every year, and in 2012 approximately 52,000 will die of the disease [190]. Although colonoscopy has significantly increased the number of cases detected at an early and manageable stage, a number of people do not have the necessary screening, in part due to cost and in part due to avoidance of the invasive procedure [191]. A molecular screening test, Cologuard, has been developed by Exact Science for the detection of pre-cancerous polyps and CRC based on DNA mutation and methylation markers combined with the immunohistochemical analysis of occult blood in stool [192–194]. However, a need still exists to identify biomarkers for the early detection of CRC in non-invasive specimens. For that reason, blood-based miRNAs are promising, and have recently been widely explored as diagnostic analytes for CRC.

In 2009, Ng et al. evaluated the expression of 95 miRNAs in plasma and CRC tissues, and found that miR-17-3p and miR-92 were elevated in both sample types [195]. Furthermore, the plasma levels of both miRNAs were reduced following curative surgery, suggesting that these markers could be used for diagnosis and monitoring of CRC. Huang and coworkers substantiated these findings by showing that plasma levels of miR-29a and miR-92a were able to distinguish healthy controls from patients with advanced CRC with an AUC of 0.844 and 0.838, respectively [196]. In the same study, the miRNAs were also used to distinguish controls from advanced adenomas, making miR-92a and miR-29a good candidates for potential early diagnostic markers for CRC. On the other hand, Wang et al. showed that miR-29a was elevated in serum of 38 CRC patients with liver metastasis when compared to CRC patients without metastasis [197], suggesting that miR-29a could be used as a marker for disease progression.

miR-21 has been demonstrated to be up-regulated in several cancer types relative to corresponding controls [198]. Levels of circulating miR-21 are also up-regulated in many cancer types including ALL, lung cancer, gastric cancer, breast cancer, pancreatic cancer, and CRC. In studies on circulating miRNAs in CRC, Pu et al. found that plasma levels of miR-21, in addition to miR-221 and miR-222, were significantly higher in 103 CRC patients relative to 37 healthy controls [199]. Interestingly, miR-221 levels were correlated with p53 as demonstrated by immunohistochemical analysis. Kanaan et al. evaluated the expression of 380 miRNAs in 30 matched normal adjacent tissues and CRC tissues and identified 19 dysregulated miRNAs [200]. Of those miRNAs, miR-21 was up-regulated in plasma of CRC patients and was able to distinguish them from healthy controls with 90% sensitivity and specificity [200].

Unlike miR-21, which has been shown to act as an oncomir, the level of miR-34 has been shown to be reduced in several cancer types, including CRC and breast cancer, delineating its function as a tumor suppressor [201].

In another large study performed on a training set of 102 plasma samples and a validation set of 156 plasma samples, Cheng and coworkers showed that miR-141 was significantly associated with stage IV colon cancer, and that the combination of miR-141 levels with CEA, a commonly used CRC marker, resulted in increased accuracy of colon cancer detection [202]. This study suggests that apart from being used as standalone markers, miRNAs may also be used to complement existing cancer markers.

5.3.2.7 Liver Cancer

Liver cancer (hepatocellular carcinoma, HCC) is the most frequent cancer type worldwide, accounting for over 750,000 new cases and 700,000 deaths every year [203]. The major etiologies for HCC include hepatitis B virus infection, alcohol abuse, and toxic chemical exposure. HCC is often diagnosed at a late stage when treatment options are limited and the prognosis is poor; early diagnosis is thus instrumental to being able to control the disease. Because HCC has been much more common in developing countries, cost-effective diagnosis and non-invasive sample collection are needed. To address this persistent need, circulating miRNAs have been evaluated for their ability to diagnose HCC.

miR-122 has been consistently identified as an important miRNA marker in liver, and is the first miRNA to be targeted in clinical trials (see the section 'miRNAs as Potential Therapeutics'). Jopling and colleagues were among the first to show that miR-122 interacts with the 5' non-coding region of hepatitis C virus, and further that miR-122 could act to facilitate replication of viral RNA [204]. Gui et al. examined the expression of miRNA using TaqMan miRNA arrays and reported that the expression of miR-885-5p was significantly higher in sera from HCC and liver cirrhosis as compared to healthy controls [205]. However, this study did not reveal any correlation between miR-885-5p expression and liver function protein. Zhang and coworkers examined miRNA expression in plasma from two mouse models (D-galactosamine and alcohol-induced liver injury) and in human plasma samples collected from 83 patients with chronic Hepatitis B viral infections, 15 patients with skeletal muscle disease, and 40 healthy volunteers. In plasma from both humans and mice, increased levels of miR-122 were associated with aminotransferase activity in the blood [206]. Interestingly, changes in miR-122 expression were specific to damage to the liver but not other organs, and most importantly, correlated with liver histologic stage. Zhang's findings suggested miR-122 as a potential diagnostic and predictive marker for viral, alcohol, and chemical-induced liver damage.

Another study examined expression of three miRNAs (miR-21, miR-122, and miR-223) in sera from 101 HCC patients and 89 healthy controls [207]. These miRNAs are commonly up-regulated in HCC tissues, and, as it turned out, they were also expressed at significantly higher levels in serum samples collected from patients with HCC or chronic hepatitis than in the controls. This observation suggests that serum levels of miR-122, miR-21, and miR-223 could be useful markers for liver injury. Substantiating these findings for miR-122, Qi et al. also found increased levels of miR-122 in HCC (N = 48) compared to controls (N = 24) [208]. However, the miRNA did not distinguish HBV patients with and without HCC [207,208]. The potential of using circulating miR-122 in HCC has been controversial, as other studies show down-regulation in HCC instead of up-regulation [209].

In an effort to identify circulating miRNAs that could distinguish HCC from chronic liver disease, Qu et al. examined the expression levels of miR-16, miR-195, and miR-199a, alone or in combination with conventional serum markers for HCC, in sera of 105 HCC patients, 107 chronic liver disease patients, and 71 healthy controls [210]. The authors found that miR-16 and miR-199a were significantly lower in HCC compared to chronic liver disease patients or controls. Although miR-16 alone was able to distinguish HCC from controls, combining miR-16 with conventional markers (such as α-fetoprotein) yielded improved sensitivity and specificity for HCC compared to either miR-16 alone or conventional markers alone [210]. Similarly to the other cancer types, circulating miR-21 has also been shown to be diagnostic for HCC [211], although this finding is controversial [212].

5.3.2.8 Prostate Cancer

Prostate cancer is second only to lung cancer in worldwide incidence, and, in the US, it is the second most common cause of cancer mortality in men aged 40 years or older. Prostate cancer in general is categorized into two forms, castrate sensitive and castrate resistant. The latter form often leads to bone metastasis and is incurable. Since prostate cancer is usually an indolent disease that predominantly occurs in an aging population, management at times is limited to watchful waiting. However, the challenge of identifying the minority of men with aggressive disease remains, as the treatment carries risks which can severely affect quality of life. Currently, prostate specific antigen (PSA) is the only approved blood-based marker for screening of prostate cancer. However, PSA is non-specific because it is also elevated in benign prostate hyperplasia (BPH), and therefore cannot be accurately used to distinguish cancer from chronic inflammation of prostate tissue. In 2012, the FDA approved the PROGENSA® PCA3 assay by Gen-Probe, Inc. The test is an RNA-based assay in which the expression levels for PCA3 and PSA are measured in the first catch urine sample following a digital-rectal examination in men aged 50 years old or older. The ratio of PCA3 to PSA RNA molecules is reported as a score that is used to guide the urologist in deciding whether a repeat biopsy is needed in men with prior negative prostate biopsies, thereby reducing the number of unnecessary biopsies. Evidence so far has also indicated that the PCA3 test is more accurate in discriminating prostate cancer than PSA alone [213,214]. These findings support the utility of RNA detection in non-invasive sample types to improve disease management. Circulating miRNAs are a non-invasive alternative biomarker that may also aid in the diagnosis of prostate cancer.

Several miRNAs including miR-34a have been shown to be dysregulated in prostate cancer tissue relative to normal [215]. In addition, miR-15a, miR-16, miR-21, miR-125b, miR-143, miR-145, miR-200 family, miR-221, miR-222, and miR-448 have been shown to influence the progression of prostate cancer (reviewed in [215]) as well as regulate key cancer genes like BCL2, WNT3A, VEGF, KRAS, TPM1, and PDCD4 (reviewed in [215]). Although this is an active field of research, this section focuses only on reviewing work on circulating miRNAs.

Analysis of serum miRNAs in 10 prostate cancer patients, patients with BPH, and healthy controls revealed that miR-26a could distinguish prostate cancer form BPH subjects [216]. Most importantly, Mahn's studies showed that miR-195 and let-7i were significantly correlated with Gleason score, and further that the levels of these circulating miRNAs were decreased after prostatectomy [216]. Circulating murine miR-141, miR-375, and miR-298 were demonstrated to be highly expressed in sera of TRAMP mice with advanced prostate cancer relative to healthy controls [217]. The human homologs of the same three miRNAs were confirmed to be up-regulated in sera of 25 men with prostate cancer relative to sera from 25 healthy controls. This confirmation of animal model results in human specimens validates the use of animal models in discovery of circulating miRNA biomarkers, and furthermore strongly suggested that miR-141 and miR-375 could be used as potential biomarkers for detection of prostate cancer. The role of circulating miR-141 and miR-375 in prostate cancer was also further confirmed by the findings of Bryant et al. using plasma derived micro-vesicles from 78 prostate cancer patients and 28 controls [218]. In that study, Bryant showed that plasma miR-141 and miR-375 were associated with metastatic prostate cancer, a finding that was also confirmed in serum from an independent patient group. An interesting finding from Bryant's studies was that urine levels of miR-107 and miR-574-3p were much higher in prostate cancer compared to controls, supporting the use of urine as a potential biomarker specimen [218]. Gonzales et al. examined the utility of plasma miR-141 as a treatment response marker in 21 prostate cancer patients and correlated the data to PSA and CTCs as well as lactate dehydrogenase [219]. Each patient had multiple serial collections of plasma during the course of the study. miR-141 was found to be correlated with clinical outcome and was highly correlated with PSA changes over time. The miR-141 and miR-375 findings were also supported by studies by Nguyen et al. who showed that serum miR-141, miR-375, and miR-378* were expressed at significantly higher levels in patients with localized prostate cancer compared those with metastatic disease [220]. Both miR-141 and miR-375 have also been shown to be overexpressed in prostate tumors compared to normal prostate tissue. Based on these reports, both miR-141 and miR-375 are good candidates for diagnosis as well as predicting disease course. Recently, Shen and coworkers examined the ability of plasma miRNA to identify prostate cancer patients with aggressive disease [221]. The study showed that miR-20a, miR-21, miR-145, and miR-221 were able to distinguish low risk from high risk prostate cancer as classified by the D'Amico system. Levels of miRs-20a and miR-21 were elevated and associated with high CAPRA scores [221], a measure used to predict recurrence.

5.4 THE CHALLENGE OF miRNAs AS BIOMARKERS

The expression of miRNA is influenced by patient-specific factors such as the specific disease condition. However, not fully understood are other factors that affect miRNA expression and detection. In general, these can be classified into pre-analytical and analytical variables. While the patient-specific variables such as smoking and alcohol cannot be controlled, pre-analytical variables such as the choice of sample type (e.g., whole blood, plasma, or serum), sample numbers, study design, sample collection method, sample stabilization and storage, and RNA extraction method can be controlled. Using the example of circulating miRNAs in lung cancer, it is important to note that, although all of the studies supported the diagnostic potential of miRNA in lung cancer, the putative miRNA biomarkers reported vary with each study even in cases when the same type of biofluid was used. A review by Brase et al. [118] discusses the putative circulating diagnostic miRNAs in different cancer types and highlights variables such as sample size, number of miRNAs examined, technology used, and normalization. All of these variables can influence the outcome, emphasizing the importance of standardization and quality-controlled methods. Further illustrating the critical importance of standardization and quality control, Kim et al. retracted an article in which they had shown that specific miRNAs are selectively destabilized depending on the adhesion status of the cells [222]. The retraction was based on findings from the same laboratory that the RNA extraction reagent biased the recovery of miRNAs with different GC content, highlighting the extraction method as an important variable that may contribute to the inconsistency of miRNA biomarker findings. Standardized, quality-controlled collection, nucleic acid extraction, and testing methods are of the utmost importance in enabling miRNA as robust and reliable biomarkers and in expediting the transition of miRNA research to clinical practice.

5.5 CONCLUSIONS

A significant number of human protein coding genes are regulated by miRNAs. miRNA-based therapies provide potentially attractive opportunities as these molecules can target multiple targets in different signaling pathways simultaneously. Furthermore, miRNAs are released into the circulation, and measurement of such species in plasma, serum, and whole blood specimens has shown early promise with respect to their acting as potential biomarkers for early disease detection, prognostic assessment, and companion diagnostic for therapeutic interventions. In the near future, large, independent, well-characterized, family and population-based case-control and additional validation studies are warranted to fully realize the potential of miRNAs as biomarkers for disease diagnosis and predictive markers for targeted therapies.

References

[1] Xie W, Ted Brown W, Denman RB. Translational regulation by non-protein-coding RNAs: different targets, common themes. Biochem Biophys Res Commun 2008;373:462–6.
[2] Xu P, Vernooy SY, Guo M, Hay BA. The Drosophila microRNA Mir-14 suppresses cell death and is required for normal fat metabolism. Curr Biol 2003;13:790–5.
[3] Brennecke J, Hipfner DR, Stark A, Russell RB, Cohen SM. Bantam encodes a developmentally regulated micro-RNA that controls cell proliferation and regulates the proapoptotic gene hid in Drosophila. Cell 2003;113:25–36.
[4] Chang S, Johnston Jr RJ, Frokjaer-Jensen C, Lockery S, Hobert O. MicroRNAs act sequentially and asymmetrically to control chemosensory laterality in the nematode. Nature 2004;430:785–9.
[5] Dostie J, Mourelatos Z, Yang M, Sharma A, Dreyfuss G. Numerous microRNPs in neuronal cells containing novel microRNAs. RNA 2003;9:180–6.
[6] Griffiths-Jones S. The microRNA Registry. Nucleic Acids Res 2004;32:D109–11.
[7] Orozco AF, Lewis DE. Flow cytometric analysis of circulating microparticles in plasma. Cytometry A 2010;77:502–14.
[8] Bentwich I, Avniel A, Karov Y, Aharonov R, Gilad S, Barad O, et al. Identification of hundreds of conserved and nonconserved human microRNAs. Nat Genet 2005;37:766–70.
[9] Croce CM. Oncogenes and cancer. N Engl J Med 2008;358:502–11.
[10] Deng S, Calin GA, Croce CM, Coukos G, Zhang L. Mechanisms of microRNA deregulation in human cancer. Cell Cycle 2008;7:2643–6.

[11] Jansen H, Reesink H, Zeuzem S, Lawitz E, Rodriguez-Torres M, Chen A. AASLD 2011.

[12] Reesink H, Janssen H, Zeuzem S. Forty seventh International Liver Congress (European Association for the Study of the Liver 2012) Barcelona. April 18–22, 2012.

[13] Bader AG. miR-34 – a microRNA replacement therapy is headed to the clinic. Front Genet 2012;3:120.

[14] Szafranska-Schwarzbach AE, Adai AT, Lee LS, Conwell DL, Andruss BF. Development of a miRNA-based diagnostic assay for pancreatic ductal adenocarcinoma. Expert Rev Mol Diagn 2011;11:249–57.

[15] Dobbin KK, Simon RM. Sample size planning for developing classifiers using high-dimensional DNA microarray data. Biostatistics 2007;8:101–17.

[16] Dobbin KK, Zhao Y, Simon RM. How large a training set is needed to develop a classifier for microarray data? Clin Cancer Res 2008;14:108–14.

[17] Tibshirani R. A simple method for assessing sample sizes in microarray experiments. BMC Bioinformatics 2006;7:106.

[18] Lee ML, Whitmore GA. Power and sample size for DNA microarray studies. Stat Med 2002;21:3543–70.

[19] Page GP, Edwards JW, Gadbury GL, Yelisetti P, Wang J, Trivedi P, et al. The PowerAtlas: a power and sample size atlas for microarray experimental design and research. BMC Bioinformatics 2006;7:84.

[20] Hess KR, Anderson K, Symmans WF, Valero V, Ibrahim N, Mejia JA, et al. Pharmacogenomic predictor of sensitivity to preoperative chemotherapy with paclitaxel and fluorouracil, doxorubicin, and cyclophosphamide in breast cancer. J Clin Oncol 2006;24:4236–44.

[21] Marchionni L, Wilson RF, Wolff AC, Marinopoulos S, Parmigiani G, Bass EB, et al. Systematic review: gene expression profiling assays in early-stage breast cancer. Ann Intern Med 2008;148:358–69.

[22] Shedden K, Taylor JM, Enkemann SA, Tsao MS, Yeatman TJ, Gerald WL, et al. Gene expression-based survival prediction in lung adenocarcinoma: a multi-site, blinded validation study. Nat Med 2008;14:822–7.

[23] Wang Y, Klijn JG, Zhang Y, Sieuwerts AM, Look MP, Yang F, et al. Gene-expression profiles to predict distant metastasis of lymph-node-negative primary breast cancer. Lancet 2005;365:671–9.

[24] Shi L, Campbell G, Jones WD, Campagne F, Wen Z, Walker SJ, et al. The MicroArray Quality Control (MAQC)-II study of common practices for the development and validation of microarray-based predictive models. Nat Biotechnol 2010;28:827–38.

[25] Golub TR, Slonim DK, Tamayo P, Huard C, Gaasenbeek M, Mesirov JP, et al. Molecular classification of cancer: class discovery and class prediction by gene expression monitoring. Science 1999;286:531–7.

[26] van 't Veer LJ, Dai H, van de Vijver MJ, He YD, Hart AA, Mao M, et al. Gene expression profiling predicts clinical outcome of breast cancer. Nature 2002;415:530–6.

[27] Beer DG, Kardia SL, Huang CC, Giordano TJ, Levin AM, Misek DE, et al. Gene-expression profiles predict survival of patients with lung adenocarcinoma. Nat Med 2002;8:816–24.

[28] Fielden MR, Adai A, Dunn II RT, Olaharski A, Searfoss G, Sina J, Predictive Safety Testing Consortium, Carcinogenicity Working Group. Development and evaluation of a genomic signature for the prediction and mechanistic assessment of nongenotoxic hepatocarcinogens in the rat. Toxicol Sci 2011;124:54–74.

[29] Dupuy A, Simon RM. Critical review of published microarray studies for cancer outcome and guidelines on statistical analysis and reporting. J Natl Cancer Inst 2007;99:147–57.

[30] Simon R. Roadmap for developing and validating therapeutically relevant genomic classifiers. J Clin Oncol 2005;23:7332–41.

[31] Simon R. Interpretation of genomic data: questions and answers. Semin Hematol 2008;45:196–204.

[32] Smyth GK, Michaud J, Scott HS. Use of within-array replicate spots for assessing differential expression in microarray experiments. Bioinformatics 2005;21:2067–75.

[33] Robinson MD, McCarthy DJ, Smyth GK. edgeR: a Bioconductor package for differential expression analysis of digital gene expression data. Bioinformatics 2010;26:139–40.

[34] J. Andre Knottnerus (Ed). The Evidence Base of Clinical Diagnosis. London: BMJ Publishing Group; 2002.

[35] Subramanian J, Simon R. What should physicians look for in evaluating prognostic gene-expression signatures? Nat Rev Clin Oncol 2010;7:327–34.

[36] Rifai N, Altman DG, Bossuyt PM. Reporting bias in diagnostic and prognostic studies: time for action. Clin Chem 2008;54:1101–3.

[37] Whiting P, Rutjes AW, Reitsma JB, Bossuyt PM, Kleijnen J. The development of QUADAS: a tool for the quality assessment of studies of diagnostic accuracy included in systematic reviews. BMC Med Res Methodol 2003;3:25.

[38] Bossuyt PM, Reitsma JB, Bruns DE, Gatsonis CA, Glasziou PP, Irwig LM, et al. Standards for Reporting of Diagnostic Accuracy. Towards complete and accurate reporting of studies of diagnostic accuracy: the STARD initiative. Standards for Reporting of Diagnostic Accuracy. Clin Chem 2003;49:1–6.

[39] Shi L, Reid LH, Jones WD, Shippy R, Warrington JA, Baker SC, et al. The microarray quality control (MAQC) project shows inter- and intraplatform reproducibility of gene expression measurements. Nat Biotechnol 2006;24:1151–61.

[40] Archer KJ, Dumur CI, Taylor GS, Chaplin MD, Guiseppi-Elie A, Grant G, et al. Application of a correlation correction factor in a microarray cross-platform reproducibility study. BMC Bioinformatics 2007;8:447.

[41] Yauk CL, Berndt ML. Review of the literature examining the correlation among DNA microarray technologies. Environ Mol Mutagen 2007;48:380–94.

[42] Fan X, Lobenhofer EK, Chen M, Shi W, Huang J, Luo J, et al. Consistency of predictive signature genes and classifiers generated using different microarray platforms. Pharmacogenomics J 2010;10:247–57.

[43] Sah S, McCall MN, Eveleigh D, Wilson M, Irizarry RA. Performance evaluation of commercial miRNA expression array platforms. BMC Res Notes 2010;3:80.

[44] Dallas PB, Gottardo NG, Firth MJ, Beesley AH, Hoffmann K, Terry PA, et al. Gene expression levels assessed by oligonucleotide microarray analysis and quantitative real-time RT-PCR – how well do they correlate? BMC Genomics 2005;6:59.

[45] Canales RD, Luo Y, Willey JC, Austermiller B, Barbacioru CC, Boysen C, et al. Evaluation of DNA microarray results with quantitative gene expression platforms. Nat Biotechnol 2006;24:1115–22.

[46] Chen Y, Gelfond JA, McManus LM, Shireman PK. Reproducibility of quantitative RT-PCR array in miRNA expression profiling and comparison with microarray analysis. BMC Genomics 2009;10:407.

[47] Gyorffy B, Molnar B, Lage H, Szallasi Z, Eklund AC. Evaluation of microarray preprocessing algorithms based on concordance with RT-PCR in clinical samples. PloS One 2009;4:e5645.

[48] Peltier HJ, Latham GJ. Normalization of microRNA expression levels in quantitative RT-PCR assays: identification of suitable reference RNA targets in normal and cancerous human solid tissues. RNA 2008;14:844–52.

[49] Mestdagh P, Van Vlierberghe P, De Weer A, Muth D, Westermann F, Speleman F, et al. A novel and universal method for microRNA RT-qPCR data normalization. Genome Biol 2009;10:R64.

[50] Latham GJ. Normalization of microRNA quantitative RT-PCR data in reduced scale experimental designs. Methods Mol Biol 2010;667:19–31.

[51] Benes V, Castoldi M. Expression profiling of microRNA using real-time quantitative PCR, how to use it and what is available. Methods 2010;50:244–9.

[52] Wylie D, Shelton J, Choudhary A, Adai AT. A novel mean-centering method for normalizing microRNA expression from high-throughput RT-qPCR data. BMC Res Notes 2011;4:555.

[53] Vandesompele J, De Preter K, Pattyn F, Poppe B, Van Roy N, De Paepe A, et al. Accurate normalization of real-time quantitative RT-PCR data by geometric averaging of multiple internal control genes. Genome Biol 2002;3(7) RESEARCH0034.

[54] Andersen CL, Jensen JL, Orntoft TF. Normalization of real-time quantitative reverse transcription-PCR data: a model-based variance estimation approach to identify genes suited for normalization, applied to bladder and colon cancer data sets. Cancer Res 2004;64:5245–50.

[55] Dunn BK, Wagner PD, Anderson D, Greenwald P. Molecular markers for early detection. Semin Oncol 2010;37:224–42.

[56] Ludwig JA, Weinstein JN. Biomarkers in cancer staging, prognosis and treatment selection. Nat Rev Cancer 2005;5:845–56.

[57] Ballehaninna UK, Chamberlain RS. The clinical utility of serum CA 19–9 in the diagnosis, prognosis and management of pancreatic adenocarcinoma: an evidence based appraisal. J Gastrointest Oncol 2012;3:105–19.

[58] Cancer Facts & Figures 2012. http://www.cancer.org/acs/groups/content/@epidemiologysurveilance/documents/document/acspc-031941.pdf

[59] Fuhrman GM, Charnsangavej C, Abbruzzese JL, Cleary KR, Martin RG, Fenoglio CJ, et al. Thin-section contrast-enhanced computed tomography accurately predicts the resectability of malignant pancreatic neoplasms. Am J Surg 1994;167:104–11; discussion 111–3.

[60] Chang KJ, Nguyen P, Erickson RA, Durbin TE, Katz KD. The clinical utility of endoscopic ultrasound-guided fine-needle aspiration in the diagnosis and staging of pancreatic carcinoma. Gastrointest Endosc 1997;45:387–93.

[61] Chaya C, Nealon WH, Bhutani MS. EUS or percutaneous CT/US-guided FNA for suspected pancreatic cancer: when tissue is the issue. Gastrointest Endosc 2006;63:976–8.

[62] Diehl SJ, Lehmann KJ, Sadick M, Lachmann R, Georgi M. Pancreatic cancer: value of dual-phase helical CT in
assessing resectability. Radiology 1998;206:373–8.

[63] Imbriaco M, Megibow AJ, Camera L, Pace L, Mainenti PP, Romano M, et al. Dual-phase versus single-phase
helical CT to detect and assess resectability of pancreatic carcinoma. Am J Roentgenol 2002;178:1473–9.

[64] Volmar KE, Vollmer RT, Jowell PS, Nelson RC, Xie HB. Pancreatic FNA in 1000 cases: a comparison of imaging modalities. Gastrointest Endosc 2005;61:854–61.

[65] Maringhini A, Ciambra M, Raimondo M, Baccelliere P, Grasso R, Dardanoni G, et al. Clinical presentation
and ultrasonography in the diagnosis of pancreatic cancer. Pancreas 1993;8:146–50.

[66] Chen Y, Zheng B, Robbins DH, Lewin DN, Mikhitarian K, Graham A, et al. Accurate discrimination of pancreatic ductal adenocarcinoma and chronic pancreatitis using multimarker expression data and samples
obtained by minimally invasive fine needle aspiration. Int J Cancer 2007;120:1511–17.

[67] Eloubeidi MA, Jhala D, Chhieng DC, Chen VK, Eltoum I, Vickers S, et al. Yield of endoscopic ultrasound-
guided fine-needle aspiration biopsy in patients with suspected pancreatic carcinoma. Cancer 2003;99:285–92.

[68] Raut CP, Grau AM, Staerkel GA, Kaw M, Tamm EP, Wolff RA, et al. Diagnostic accuracy of endoscopic
ultrasound-guided fine-needle aspiration in patients with presumed pancreatic cancer. J Gastrointest Surg
2003;7:118–26; discussion 127–8.

[69] Agarwal B, Abu-Hamda E, Molke KL, Correa AM, Ho L. Endoscopic ultrasound-guided fine needle aspiration and multidetector spiral CT in the diagnosis of pancreatic cancer. Am J Gastroenterol 2004;99:844–50.

[70] Taylor B. Carcinoma of the head of the pancreas versus chronic pancreatitis: diagnostic dilemma with significant consequences. World J Surg 2003;27:1249–57.

[71] Graham RA, Bankoff M, Hediger R, Shaker HZ, Reinhold RB. Fine-needle aspiration biopsy of pancreatic
ductal adenocarcinoma: loss of diagnostic accuracy with small tumors. J Surg Oncol 1994;55:92–4.

[72] Jhala NC, Jhala DN, Chhieng DC, Eloubeidi MA, Eltoum IA. Endoscopic ultrasound-guided fine-needle aspiration. A cytopathologist's perspective. Am J Clin Pathol 2003;120:351–67.

[73] Zalatnai A. Pathologic diagnosis of pancreatic cancer – facts, pitfalls, challenges. Orv Hetil 2001;142:
1885–90.

[74] Shin HJ, Lahoti S, Sneige N. Endoscopic ultrasound-guided fine-needle aspiration in 179 cases: the M. D.
Anderson Cancer Center experience. Cancer 2002;96:174–80.

[75] Chang KJ, Katz KD, Durbin TE, Erickson RA, Butler JA, Lin F, et al. Endoscopic ultrasound-guided fine-needle aspiration. Gastrointest Endosc 1994;40:694–9.

[76] Laurell H, Bouisson M, Berthelemy P, Rochaix P, Dejean S, Besse P, et al. Identification of biomarkers of
human pancreatic adenocarcinomas by expression profiling and validation with gene expression analysis in
endoscopic ultrasound-guided fine needle aspiration samples. World J Gastroenterol 2006;12:3344–51.

[77] van Gulik TM, Reeders JW, Bosma A, Moojen TM, Smits NJ, Allema JH, et al. Incidence and clinical findings of benign, inflammatory disease in patients resected for presumed pancreatic head cancer. Gastrointest
Endosc 1997;46:417–23.

[78] Ardengh JC, Lopes CV, Campos AD, Pereira de Lima LF, Venco F, Modena JL. Endoscopic ultrasound and
fine needle aspiration in chronic pancreatitis: differential diagnosis between pseudotumoral masses and pancreatic cancer. JOP 2007;8:413–21.

[79] Gudjonsson B, Livstone EM, Spiro HM. Cancer of the pancreas: diagnostic accuracy and survival statistics.
Cancer 1978;42:2494–506.

[80] Erickson RA. Pancreatic cancer. In: Balducci L, Talavera F, Movsas B, McKenna R, Macdonald JS, editors.
eMedicine from WebMD vol. 2007; 2005.

[81] Yachida S, Jones S, Bozic I, Antal T, Leary R, Fu B, et al. Distant metastasis occurs late during the genetic evolution of pancreatic cancer. Nature 2010;467:1114–17.

[82] Howlader N, Noone A, Krapcho M, Neyman N, Aminou R, Waldron W, et al. SEER Cancer Statistics Review,
1975–2008 2011 edit. SEER. 2011.

[83] Gironella M, Seux M, Xie MJ, Cano C, Tomasini R, Gommeaux J, et al. Tumor protein 53-induced nuclear
protein 1 expression is repressed by miR-155, and its restoration inhibits pancreatic tumor development. Proc
Natl Acad Sci USA 2007;104:16170–16175.

[84] Chang TC, Wentzel EA, Kent OA, Ramachandran K, Mullendore M, Lee KH, et al. Transactivation of miR-
34a by p53 broadly influences gene expression and promotes apoptosis. Mol Cell 2007;26:745–52.

[85] Szafranska AE, Davison TS, John J, Cannon T, Sipos B, Maghnouj A, et al. MicroRNA expression alterations are linked to tumorigenesis and non-neoplastic processes in pancreatic ductal adenocarcinoma. Oncogene 2007;26:4442–52.

[86] Bloomston M, Frankel WL, Petrocca F, Volinia S, Alder H, Hagan JP, et al. MicroRNA expression patterns to differentiate pancreatic adenocarcinoma from normal pancreas and chronic pancreatitis. JAMA 2007;297:1901–8.

[87] Zhang Y, Li M, Wang H, Fisher WE, Lin PH, Yao Q, et al. Profiling of 95 microRNAs in pancreatic cancer cell lines and surgical specimens by real-time PCR analysis. World J Surg 2009;33:698–709.

[88] Dillhoff M, Liu J, Frankel W, Croce C, Bloomston M. MicroRNA-21 is overexpressed in pancreatic cancer and a potential predictor of survival. J Gastrointest Surg 2008;12:2171–6.

[89] Lee EJ, Gusev Y, Jiang J, Nuovo GJ, Lerner MR, Frankel WL, et al. Expression profiling identifies microRNA signature in pancreatic cancer. Int J Cancer 2007;120:1046–54.

[90] Munding JB, Adai AT, Maghnouj A, Urbanik A, Zollner H, Liffers ST, et al. Global microRNA expression profiling of microdissected tissues identifies miR-135b as a novel biomarker for pancreatic ductal adenocarcinoma. Int J Cancer 2012;131(2):E86–95.

[91] Greither T, Grochola LF, Udelnow A, Lautenschlager C, Wurl P, Taubert H. Elevated expression of microRNAs 155, 203, 210 and 222 in pancreatic tumors is associated with poorer survival. Int J Cancer 2010;126:73–80.

[92] Szafranska AE, Doleshal M, Edmunds HS, Gordon S, Luttges J, Munding JB, et al. Analysis of microRNAs in pancreatic fine-needle aspirates can classify benign and malignant tissues. Clin Chem 2008;54:1716–24.

[93] Yu J, Li A, Hong SM, Hruban RH, Goggins M. MicroRNA alterations of pancreatic intraepithelial neoplasias. Clin Cancer Res 2012;18:981–92.

[94] Hanoun N, Delpu Y, Suriawinata AA, Bournet B, Bureau C, Selves J, et al. The silencing of microRNA 148a production by DNA hypermethylation is an early event in pancreatic carcinogenesis. Clin Chem 2010;56:1107–18.

[95] Matthaei H, Maitra A. Precursor lesions of pancreatic cancer, Fitzgerald RC, editor. Pre-invasive disease: pathogenesis and clinical management. New York: Springer; 2011, pp. 395–420.

[96] Laffan TA, Horton KM, Klein AP, Berlanstein B, Siegelman SS, Kawamoto S, et al. Prevalence of unsuspected pancreatic cysts on MDCT. Am J Roentgenol 2008;191:802–7.

[97] Visser BC, Yeh BM, Qayyum A, Way LW, McCulloch CE, Coakley FV. Characterization of cystic pancreatic masses: relative accuracy of CT and MRI. AJR Am J Roentgenol 2007;189:648–56.

[98] Brugge WR, Lewandrowski K, Lee-Lewandrowski E, Centeno BA, Szydlo T, Regan S, et al. Diagnosis of pancreatic cystic neoplasms: a report of the cooperative pancreatic cyst study. Gastroenterology 2004;126:1330–6.

[99] Gerke H, Jaffe TA, Mitchell RM, Byrne MF, Stiffler HL, Branch MS, et al. Endoscopic ultrasound and computer tomography are inaccurate methods of classifying cystic pancreatic lesions. Dig Liver Dis 2006;38:39–44.

[100] Ahmad NA, Kochman ML, Brensinger C, Brugge WR, Faigel DO, Gress FG, et al. Interobserver agreement among endosonographers for the diagnosis of neoplastic versus non-neoplastic pancreatic cystic lesions. Gastrointest Endosc 2003;58:59–64.

[101] Lai R, Stanley MW, Bardales R, Linzie B, Mallery S. Endoscopic ultrasound-guided pancreatic duct aspiration: diagnostic yield and safety. Endoscopy 2002;34:715–20.

[102] Maire F, Couvelard A, Hammel P, Ponsot P, Palazzo L, Aubert A, et al. Intraductal papillary mucinous tumors of the pancreas: the preoperative value of cytologic and histopathologic diagnosis. Gastrointest Endosc 2003;58:701–6.

[103] Frossard JL, Amouyal P, Amouyal G, Palazzo L, Amaris J, Soldan M, et al. Performance of endosonography-guided fine needle aspiration and biopsy in the diagnosis of pancreatic cystic lesions. Am J Gastroenterol 2003;98:1516–24.

[104] van der Waaij LA, van Dullemen HM, Porte RJ. Cyst fluid analysis in the differential diagnosis of pancreatic cystic lesions: a pooled analysis. Gastrointest Endosc 2005;62:383–9.

[105] Kucera S, Centeno BA, Springett G, Malafa MP, Chen YA, Weber J, et al. Cyst fluid carcinoembryonic antigen level is not predictive of invasive cancer in patients with intraductal papillary mucinous neoplasm of the pancreas. JOP 2012;13:409–13.

[106] Khalid A, Zahid M, Finkelstein SD, LeBlanc JK, Kaushik N, Ahmad N, et al. Pancreatic cyst fluid DNA analysis in evaluating pancreatic cysts: a report of the PANDA study. Gastrointest Endosc 2009;69:1095–102.

[107] Wu J, Matthaei H, Maitra A, Dal Molin M, Wood LD, Eshleman JR, et al. Recurrent GNAS mutations define an unexpected pathway for pancreatic cyst development. Sci Transl Med 2011;3:92ra66.

[108] Habbe N, Koorstra JB, Mendell JT, Offerhaus GJ, Ryu JK, Feldmann G, et al. MicroRNA miR-155 is a biomarker of early pancreatic neoplasia. Cancer Biol Ther 2009;8:340–6.

[109] Ryu JK, Matthaei H, Dal Molin M, Hong SM, Canto MI, Schulick RD, et al. Elevated microRNA miR-21 levels in pancreatic cyst fluid are predictive of mucinous precursor lesions of ductal adenocarcinoma. Pancreatology 2011;11:343–50.

[110] Matthaei H, Wylie D, Lloyd MB, Dal Molin M, Kemppainen J, Mayo SC, et al. miRNA biomarkers in cyst fluid augment the diagnosis and management of pancreatic cysts. Clin Cancer Res 2012;18:4713–24.

[111] Giovannetti E, Funel N, Peters GJ, Del Chiaro M, Erozenci LA, Vasile E, et al. MicroRNA-21 in pancreatic cancer: correlation with clinical outcome and pharmacologic aspects underlying its role in the modulation of gemcitabine activity. Cancer Res 2010;70:4528–38.

[112] Tavano F, di Mola FF, Piepoli A, Panza A, Copetti M, Burbaci FP, et al. Changes in miR-143 and miR-21 expression and clinicopathological correlations in pancreatic cancers. Pancreas 2012;41:1280–4.

[113] Kong X, Du Y, Wang G, Gao J, Gong Y, Li L, et al. Detection of differentially expressed microRNAs in serum of pancreatic ductal adenocarcinoma patients: miR-196a could be a potential marker for poor prognosis. Dig Dis Sci 2011;56:602–9.

[114] Preis M, Gardner TB, Gordon SR, Pipas JM, Mackenzie TA, Klein EE, et al. MicroRNA-10b expression correlates with response to neoadjuvant therapy and survival in pancreatic ductal adenocarcinoma. Clin Cancer Res 2011;17:5812–21.

[115] Murray CJ, Lopez AD. Alternative projections of mortality and disability by cause 1990–2020: global burden of disease study. Lancet 1997;349:1498–504.

[116] Lawrie CH, Gal S, Dunlop HM, Pushkaran B, Liggins AP, Pulford K, et al. Detection of elevated levels of tumor-associated microRNAs in serum of patients with diffuse large B-cell lymphoma. Br J Haematol 2008;141:672–5.

[117] Mitchell PS, Parkin RK, Kroh EM, Fritz BR, Wyman SK, Pogosova-Agadjanyan EL, et al. Circulating microRNAs as stable blood-based markers for cancer detection. Proc Natl Acad Sci USA 2008;105:10513–18.

[118] Brase JC, Wuttig D, Kuner R, Sultmann H. Serum microRNAs as non-invasive biomarkers for cancer. Mol Cancer 2010;9:306.

[119] Hunter MP, Ismail N, Zhang X, Aguda BD, Lee EJ, Yu L, et al. Detection of microRNA expression in human peripheral blood microvesicles. PLoS One 2008;3:e3694.

[120] Michael A, Bajracharya SD, Yuen PS, Zhou H, Star RA, Illei GG, et al. Exosomes from human saliva as a source of microRNA biomarkers. Oral Dis 2010;16:34–8.

[121] Dimov I, Jankovic Velickovic L, Stefanovic V. Urinary exosomes. ScientificWorldJournal 2009;9:1107–18.

[122] Skog J, Wurdinger T, van Rijn S, Meijer DH, Gainche L, Sena-Esteves M, et al. Glioblastoma microvesicles transport RNA and proteins that promote tumor growth and provide diagnostic biomarkers. Nat Cell Biol 2008;10:1470–6.

[123] Weickmann JL, Glitz DG. Human ribonucleases. Quantitation of pancreatic-like enzymes in serum, urine, and organ preparations. J Biol Chem 1982;257:8705–10.

[124] Gibbings DJ, Ciaudo C, Erhardt M, Voinnet O. Multivesicular bodies associate with components of miRNA effector complexes and modulate miRNA activity. Nat Cell Biol 2009;11:1143–9.

[125] Valadi H, Ekstrom K, Bossios A, Sjostrand M, Lee JJ, Lotvall JO. Exosome-mediated transfer of mRNAs and microRNAs is a novel mechanism of genetic exchange between cells. Nat Cell Biol 2007;9:654–9.

[126] Pegtel DM, Cosmopoulos K, Thorley-Lawson DA, van Eijndhoven MA, Hopmans ES, Lindenberg JL, et al. Functional delivery of viral miRNAs via exosomes. Proc Natl Acad Sci USA 2010;107:6328–33.

[127] Calin GA, Ferracin M, Cimmino A, Di Leva G, Shimizu M, Wojcik SE, et al. A MicroRNA signature associated with prognosis and progression in chronic lymphocytic leukemia. N Engl J Med 2005;353:1793–801.

[128] Calin GA, Dumitru CD, Shimizu M, Bichi R, Zupo S, Noch E, et al. Frequent deletions and down-regulation of micro-RNA genes miR15 and miR16 at 13q14 in chronic lymphocytic leukemia. Proc Natl Acad Sci USA 2002;99:15524–9.

[129] Calin GA, Liu CG, Sevignani C, Ferracin M, Felli N, Dumitru CD, et al. MicroRNA profiling reveals distinct signatures in B cell chronic lymphocytic leukemias. Proc Natl Acad Sci USA 2004;101:11755–60.

[130] Pekarsky Y, Santanam U, Cimmino A, Palamarchuk A, Efanov A, Maximov V, et al. TCL1 expression in chronic lymphocytic leukemia is regulated by miR-29 and miR-181. Cancer Res 2006;66:11590–3.

[131] Tanaka M, Oikawa K, Takanashi M, Kudo M, Ohyashiki J, Ohyashiki K, et al. Down-regulation of miR-92 in human plasma is a novel marker for acute leukemia patients. PLoS One 2009;4:e5532.

[132] Costinean S, Zanesi N, Pekarsky Y, Tili E, Volinia S, Heerema N, et al. Pre-B cell proliferation and lymphoblastic leukemia/high-grade lymphoma in E(mu)-miR155 transgenic mice. Proc Natl Acad Sci USA 2006;103:7024–9.

[133] Mi S, Lu J, Sun M, Li Z, Zhang H, Neilly MB, et al. MicroRNA expression signatures accurately discriminate acute lymphoblastic leukemia from acute myeloid leukemia. Proc Natl Acad Sci USA 2007;104:19971–6.

[134] Schotte D, Chau JC, Sylvester G, Liu G, Chen C, van der Velden VH, et al. Identification of new microRNA genes and aberrant microRNA profiles in childhood acute lymphoblastic leukemia. Leukemia 2009;23:313–22.

[135] Wang J, Chen J, Chang P, LeBlanc A, Li D, Abbruzzesse JL, et al. MicroRNAs in plasma of pancreatic ductal adenocarcinoma patients as novel blood-based biomarkers of disease. Cancer Prev Res (Phila) 2009;2:807–13.

[136] Liu J, Gao J, Du Y, Li Z, Ren Y, Gu J, et al. Combination of plasma microRNAs with serum CA19–9 for early detection of pancreatic cancer. Int J Cancer 2012;131:683–91.

[137] Liu R, Chen X, Du Y, Yao W, Shen L, Wang C, et al. Serum microRNA expression profile as a biomarker in the diagnosis and prognosis of pancreatic cancer. Clin Chem 2012;58:610–18.

[138] Ren Y, Gao J, Liu JQ, Wang XW, Gu JJ, Huang HJ, et al. Differential signature of fecal microRNAs in patients with pancreatic cancer. Mol Med Report 2012;6:201–9.

[139] Reddy C, Chilla D, Boltax J. Lung cancer screening: a review of available data and current guidelines. Hosp Pract (Minneap) 2011;39:107–12.

[140] Smith RA, Duffy SW, Tabar L. Breast cancer screening: the evolving evidence. Oncology (Williston Park) 2012;26:471–5, 479–81, 485–6.

[141] Burt RW. Colorectal cancer screening. Curr Opin Gastroenterol 2010;26:466–70.

[142] Kramer BS, Berg CD, Aberle DR, Prorok PC. Lung cancer screening with low-dose helical CT: results from the national lung screening trial (NLST). J Med Screen 2011;18:109–11.

[143] Osada H, Takahashi T. let-7 and miR-17–92: small-sized major players in lung cancer development. Cancer Sci 2011;102:9–17.

[144] Dziadziuszko R, Hirsch FR. Advances in genomic and proteomic studies of non-small-cell lung cancer: clinical and translational research perspective. Clin Lung Cancer 2008;9:78–84.

[145] Sung HJ, Cho JY. Biomarkers for the lung cancer diagnosis and their advances in proteomics. BMB Rep 2008;41:615–25.

[146] Shen J, Todd NW, Zhang H, Yu L, Lingxiao X, Mei Y, et al. Plasma microRNAs as potential biomarkers for non-small-cell lung cancer. Lab Invest 2011;91:579–87.

[147] Keller A, Leidinger P, Borries A, Wendschlag A, Wucherpfennig F, Scheffler M, et al. miRNAs in lung cancer–studying complex fingerprints in patient's blood cells by microarray experiments. BMC Cancer 2009;9:353.

[148] Foss KM, Sima C, Ugolini D, Neri M, Allen KE, Weiss GJ. miR-1254 and miR-574–5p: serum-based microRNA biomarkers for early-stage non-small cell lung cancer. J Thorac Oncol 2011;6:482–8.

[149] Chen X, Ba Y, Ma L, Cai X, Yin Y, Wang K, et al. Characterization of microRNAs in serum: a novel class of biomarkers for diagnosis of cancer and other diseases. Cell Res 2008;18:997–1006.

[150] Rabinowits G, Gercel-Taylor C, Day JM, Taylor DD, Kloecker GH. Exosomal microRNA: a diagnostic marker for lung cancer. Clin Lung Cancer 2009;10:42–6.

[151] Hennessey PT, Sanford T, Choudhary A, Mydlarz WW, Brown D, Adai AT, et al. Serum microRNA biomarkers for detection of non-small cell lung cancer. PLoS One 2012;7:e32307.

[152] Patnaik SK, Yendamuri S, Kannisto E, Kucharczuk JC, Singhal S, Vachani A. MicroRNA expression profiles of whole blood in lung adenocarcinoma. PLoS One 2012;7:e46045.

[153] Wang T, Lv M, Shen S, Zhou S, Wang P, Chen Y, et al. Cell-Free microRNA expression profiles in malignant effusion associated with patient survival in non-small cell lung cancer. PLoS One 2012;7:e43268.

[154] Yuxia M, Zhennan T, Wei Z. Circulating miR-125b is a novel biomarker for screening non-small-cell lung cancer and predicts poor prognosis. J Cancer Res Clin Oncol 2012;138(12):2045–50.

[155] Le HB, Zhu WY, Chen DD, He JY, Huang YY, Liu XG, et al. Evaluation of dynamic change of serum miR-21 and miR-24 in pre- and post-operative lung carcinoma patients. Med Oncol 2012;29(5):3190–7.

[156] Zheng D, Haddadin S, Wang Y, Gu LQ, Perry MC, Freter CE, et al. Plasma microRNAs as novel biomarkers for early detection of lung cancer. Int J Clin Exp Pathol 2011;4:575–86.

[157] Shen J, Liu Z, Todd NW, Zhang H, Liao J, Yu L, et al. Diagnosis of lung cancer in individuals with solitary pulmonary nodules by plasma microRNA biomarkers. BMC Cancer 2011;11:374.

[158] Roth C, Kasimir-Bauer S, Pantel K, Schwarzenbach H. Screening for circulating nucleic acids and caspase activity in the peripheral blood as potential diagnostic tools in lung cancer. Mol Oncol 2011;5:281–91.

[159] Xie L, Chen X, Wang L, Qian X, Wang T, Wei J, et al. Cell-free miRNAs may indicate diagnosis and docetaxel sensitivity of tumor cells in malignant effusions. BMC Cancer 2010;10:591.

[160] Wei J, Gao W, Zhu CJ, Liu YQ, Mei Z, Cheng T, et al. Identification of plasma microRNA-21 as a biomarker for early detection and chemosensitivity of non-small cell lung cancer. Chin J Cancer 2011;30:407–14.

[161] Silva J, Garcia V, Zaballos A, Provencio M, Lombardia L, Almonacid L, et al. Vesicle-related microRNAs in plasma of nonsmall cell lung cancer patients and correlation with survival. Eur Respir J 2011;37:617–23.

[162] Jemal A, Siegel R, Ward E, Hao Y, Xu J, Thun MJ. Cancer statistics. CA Cancer J Clin 2009;59:225–49.

[163] Weir HK, Thun MJ, Hankey BF, Ries LA, Howe HL, Wingo PA, et al. Annual report to the nation on the status of cancer, 1975–2000, featuring the uses of surveillance data for cancer prevention and control. J Natl Cancer Inst 2003;95:1276–99.

[164] Jemal A, Siegel R, Ward E, Murray T, Xu J, Thun MJ. Cancer statistics. CA Cancer J Clin 2007;57:43–66.

[165] Harris L, Fritsche H, Mennel R, Norton L, Ravdin P, Taube S, et al. American Society of Clinical Oncology 2007 update of recommendations for the use of tumor markers in breast cancer. J Clin Oncol 2007;25:5287–312.

[166] Schrauder MG, Strick R, Schulz-Wendtland R, Strissel PL, Kahmann L, Loehberg CR, et al. Circulating micro-RNAs as potential blood-based markers for early stage breast cancer detection. PLoS One 2012;7:e29770.

[167] Zhao R, Wu J, Jia W, Gong C, Yu F, Ren Z, et al. Plasma miR-221 as a predictive biomarker for chemore-sistance in breast cancer patients who previously received neoadjuvant chemotherapy. Onkologie 2011;34:675–80.

[168] Schwarzenbach H, Milde-Langosch K, Steinbach B, Muller V, Pantel K. Diagnostic potential of PTEN-targeting miR-214 in the blood of breast cancer patients. Breast Cancer Res Treat 2012;134:933–41.

[169] Xiong X, Ren HZ, Li MH, Mei JH, Wen JF, Zheng CL. Down-regulated miRNA-214 induces a cell cycle G1 arrest in gastric cancer cells by up-regulating the PTEN protein. Pathol Oncol Res 2011;17:931–7.

[170] van Schooneveld E, Wouters MC, Van der Auwera I, Peeters DJ, Wildiers H, Van Dam PA, et al. Expression profiling of cancerous and normal breast tissues identifies microRNAs that are differentially expressed in serum from patients with (metastatic) breast cancer and healthy volunteers. Breast Cancer Res 2012;14:R34.

[171] Jung EJ, Santarpia L, Kim J, Esteva FJ, Moretti E, Buzdar AU, et al. Plasma microRNA 210 levels correlate with sensitivity to trastuzumab and tumor presence in breast cancer patients. Cancer 2012;118:2603–14.

[172] Wu X, Somlo G, Yu Y, Palomares MR, Li AX, Zhou W, et al. De novo sequencing of circulating miRNAs identifies novel markers predicting clinical outcome of locally advanced breast cancer. J Transl Med 2012;10:42.

[173] Wang H, Tan G, Dong L, Cheng L, Li K, Wang Z, et al. Circulating MiR-125b as a marker predicting chemore-sistance in breast cancer. PLoS One 2012;7:e34210.

[174] Cuk K, Zucknick M, Heil J, Madhavan D, Schott S, Turchinovich A, et al. Circulating microRNAs in plasma as early detection markers for breast cancer. Int J Cancer 2013;132(7):1602–12.

[175] Madhavan D, Zucknick M, Wallwiener M, Cuk K, Modugno C, Scharpff M, et al. Circulating microRNAs as surrogate markers for circulating tumor cells and prognostic markers in metastatic breast cancer. Clin Cancer Res 2012;18(21):5972–82.

[176] Si H, Sun X, Chen Y, Cao Y, Chen S, Wang H, et al. Circulating microRNA-92a and microRNA-21 as novel minimally invasive biomarkers for primary breast cancer. J Cancer Res Clin Oncol 2012;139(2):223–9.

[177] Ferlay J, Shin HR, Bray F, Forman D, Mathers C, Parkin DM. Estimates of worldwide burden of cancer in 2008: GLOBOCAN 2008. Int J Cancer 2010;127:2893–917.

[178] Song JH, Meltzer SJ. MicroRNAs in pathogenesis, diagnosis, and treatment of gastroesophageal cancers. Gastroenterology 2012;143:35–47, e2.

[179] Ueda T, Volinia S, Okumura H, Shimizu M, Taccioli C, Rossi S, et al. Relation between microRNA expression and progression and prognosis of gastric cancer: a microRNA expression analysis. Lancet Oncol 2010;11:136–46.

[180] Valladares-Ayerbes M, Reboredo M, Medina-Villaamil V, Iglesias-Diaz P, Lorenzo-Patino MJ, Haz M, et al. Circulating miR-200c as a diagnostic and prognostic biomarker for gastric cancer. J Transl Med 2012;10:186.

[181] Tsai KW, Liao YL, Wu CW, Hu LY, Li SC, Chan WC, et al. Aberrant expression of miR-196a in gastric cancers and correlation with recurrence. Genes Chromosomes Cancer 2012;51:394–401.

[182] Wang M, Gu H, Wang S, Qian H, Zhu W, Zhang L, et al. Circulating miR-17–5p and miR-20a: molecular markers for gastric cancer. Mol Med Report 2012;5:1514–20.

[183] Liu H, Zhu L, Liu B, Yang L, Meng X, Zhang W, et al. Genome-wide microRNA profiles identify miR-378 as a serum biomarker for early detection of gastric cancer. Cancer Lett 2012;316:196–203.

[184] Zhu S, Si ML, Wu H, Mo YY. MicroRNA-21 targets the tumor suppressor gene tropomyosin 1 (TPM1). J Biol Chem 2007;282:14328–36.

[185] Meng F, Henson R, Wehbe-Janek H, Ghoshal K, Jacob ST, Patel T. MicroRNA-21 regulates expression of the PTEN tumor suppressor gene in human hepatocellular cancer. Gastroenterology 2007;133:647–58.

[186] Asangani IA, Rasheed SA, Nikolova DA, Leupold JH, Colburn NH, Post S, et al. MicroRNA-21 (miR-21) post-transcriptionally downregulates tumor suppressor Pdcd4 and stimulates invasion, intravasation and metastasis in colorectal cancer. Oncogene 2008;27:2128–36.

[187] Zheng Y, Cui L, Sun W, Zhou H, Yuan X, Huo M, et al. MicroRNA-21 is a new marker of circulating tumor cells in gastric cancer patients. Cancer Biomark 2011;10:71–7.

[188] Zhou H, Guo JM, Lou YR, Zhang XJ, Zhong FD, Jiang Z, et al. Detection of circulating tumor cells in peripheral blood from patients with gastric cancer using microRNA as a marker. J Mol Med (Berl) 2010;88:709–17.

[189] Tsujiura M, Ichikawa D, Komatsu S, Shiozaki A, Takeshita H, Kosuga T, et al. Circulating microRNAs in plasma of patients with gastric cancers. Br J Cancer 2010;102:1174–9.

[190] Surveillance, Epidemiology, and End Results (SEER) Program, <www.seer.cancer.gov>; 2012 [accessed 11.01.13].

[191] Anderson JC, Fortinsky RH, Kleppinger A, Merz-Beyus AB, Huntington III CG, Lagarde S. Predictors of compliance with free endoscopic colorectal cancer screening in uninsured adults. J Gen Intern Med 2011;26:875–80.

[192] Ahlquist DA, Taylor WR, Mahoney DW, Zou H, Domanico M, Thibodeau SN, et al. The stool DNA test is more accurate than the plasma septin 9 test in detecting colorectal neoplasia. Clin Gastroenterol Hepatol 2012;10: 272–7, e1.

[193] Berger BM, Ahlquist DA. Stool DNA screening for colorectal neoplasia: biological and technical basis for high detection rates. Pathology 2012;44:80–8.

[194] Zou H, Taylor WR, Harrington JJ, Hussain FT, Cao X, Loprinzi CL, et al. High detection rates of colorectal neoplasia by stool DNA testing with a novel digital melt curve assay. Gastroenterology 2009;136:459–70.

[195] Ng EK, Chong WW, Jin H, Lam EK, Shin VY, Yu J, et al. Differential expression of microRNAs in plasma of patients with colorectal cancer: a potential marker for colorectal cancer screening. Gut 2009;58:1375–81.

[196] Huang Z, Huang D, Ni S, Peng Z, Sheng W, Du X. Plasma microRNAs are promising novel biomarkers for early detection of colorectal cancer. Int J Cancer 2010;127:118–26.

[197] Wang LG, Gu J. Serum microRNA-29a is a promising novel marker for early detection of colorectal liver metastasis. Cancer Epidemiol 2012;36:e61–7.

[198] Volinia S, Calin GA, Liu CG, Ambs S, Cimmino A, Petrocca F, et al. A microRNA expression signature of human solid tumors defines cancer gene targets. Proc Natl Acad Sci USA 2006;103:2257–61.

[199] Pu XX, Huang GL, Guo HQ, Guo CC, Li H, Ye S, et al. Circulating miR-221 directly amplified from plasma is a potential diagnostic and prognostic marker of colorectal cancer and is correlated with p53 expression. J Gastroenterol Hepatol 2010;25:1674–80.

[200] Kanaan Z, Rai SN, Eichenberger MR, Roberts H, Keskey B, Pan J, et al. Plasma miR-21: a potential diagnostic marker of colorectal cancer. Ann Surg 2012;256:544–51.

[201] Nugent M, Miller N, Kerin MJ. Circulating miR-34a levels are reduced in colorectal cancer. J Surg Oncol 2012;106(8):947–52.

[202] Cheng H, Zhang L, Cogdell DE, Zheng H, Schetter AJ, Nykter M, et al. Circulating plasma MiR-141 is a novel biomarker for metastatic colon cancer and predicts poor prognosis. PLoS One 2011;6:e17745.

[203] El-Serag HB, Rudolph KL. Hepatocellular carcinoma: epidemiology and molecular carcinogenesis. Gastroenterology 2007;132:2557–76.

[204] Jopling CL, Yi M, Lancaster AM, Lemon SM, Sarnow P. Modulation of hepatitis C virus RNA abundance by a liver-specific MicroRNA. Science 2005;309:1577–81.

[205] Gui J, Tian Y, Wen X, Zhang W, Zhang P, Gao J, et al. Serum microRNA characterization identifies miR-885-5p as a potential marker for detecting liver pathologies. Clin Sci (Lond) 2011;120:183–93.

[206] Zhang Y, Jia Y, Zheng R, Guo Y, Wang Y, Guo H, et al. Plasma microRNA-122 as a biomarker for viral-, alcohol-, and chemical-related hepatic diseases. Clin Chem 2010;56:1830–8.

[207] Xu J, Wu C, Che X, Wang L, Yu D, Zhang T, et al. Circulating microRNAs, miR-21, miR-122, and miR-223, in patients with hepatocellular carcinoma or chronic hepatitis. Mol Carcinog 2011;50:136–42.

[208] Qi P, Cheng SQ, Wang H, Li N, Chen YF, Gao CF. Serum microRNAs as biomarkers for hepatocellular carcinoma in Chinese patients with chronic hepatitis B virus infection. PLoS One 2011;6:e28486.

[209] Ding X, Ding J, Ning J, Yi F, Chen J, Zhao D, et al. Circulating microRNA-122 as a potential biomarker for liver injury. Mol Med Report 2012;5:1428–32.

[210] Qu KZ, Zhang K, Li H, Afdhal NH, Albitar M. Circulating microRNAs as biomarkers for hepatocellular carcinoma. J Clin Gastroenterol 2011;45:355–60.

[211] Tomimaru Y, Eguchi H, Nagano H, Wada H, Kobayashi S, Marubashi S, et al. Circulating microRNA-21 as a novel biomarker for hepatocellular carcinoma. J Hepatol 2012;56:167–75.

[212] Cermelli S, Ruggieri A, Marrero JA, Ioannou GN, Beretta L. Circulating microRNAs in patients with chronic hepatitis C and non-alcoholic fatty liver disease. PLoS One 2011;6:e23937.

[213] Ferro M, Bruzzese D, Perdona S, Mazzarella C, Marino A, Sorrentino A, et al. Predicting prostate biopsy outcome: prostate health index (phi) and prostate cancer antigen 3 (PCA3) are useful biomarkers. Clin Chim Acta 2012;413:1274–8.

[214] Ramos CG, Valdevenito R, Vergara I, Anabalon P, Sanchez C, Fulla J. PCA3 sensitivity and specificity for prostate cancer detection in patients with abnormal PSA and/or suspicious digital rectal examination. First Latin American experience. Urol Oncol 2012; <http://dx.doi.org/10.1016/j.urolonc.2012.05.002>.

[215] Hassan O, Ahmad A, Sethi S, Sarkar FH. Recent updates on the role of microRNAs in prostate cancer. J Hematol Oncol 2012;5:9.

[216] Mahn R, Heukamp LC, Rogenhofer S, von Ruecker A, Muller SC, Ellinger J. Circulating microRNAs (miRNA) in serum of patients with prostate cancer. Urology 2011;77: 1265, e9–16.

[217] Selth LA, Townley S, Gillis JL, Ochnik AM, Murti K, Macfarlane RJ, et al. Discovery of circulating microRNAs associated with human prostate cancer using a mouse model of disease. Int J Cancer 2012;131:652–61.

[218] Bryant RJ, Pawlowski T, Catto JW, Marsden G, Vessella RL, Rhees B, et al. Changes in circulating microRNA levels associated with prostate cancer. Br J Cancer 2012;106:768–74.

[219] Gonzales JC, Fink LM, Goodman Jr. OB, Symanowski JT, Vogelzang NJ, Ward DC. Comparison of circulating MicroRNA 141 to circulating tumor cells, lactate dehydrogenase, and prostate-specific antigen for determining treatment response in patients with metastatic prostate cancer. Clin Genitourin Cancer 2011;9:39–45.

[220] Nguyen HC, Xie W, Yang M, Hsieh CL, Drouin S, Lee GS, et al. Expression differences of circulating microRNAs in metastatic castration resistant prostate cancer and low-risk, localized prostate cancer. Prostate 2013 Mar;73(4):346–54.

[221] Shen J, Hruby GW, McKiernan JM, Gurvich I, Lipsky MJ, Benson MC, et al. Dysregulation of circulating microRNAs and prediction of aggressive prostate cancer. Prostate 2012;72:1469–77.

[222] Kim YK, Yeo J, Ha M, Kim B, Kim VN. Cell adhesion-dependent control of microRNA decay. Mol Cell 2011;43:1005–14.

Toxicogenomics – A Drug Development Perspective

Yuping Wang[1], Jurgen Borlak[2], Weida Tong[1]

[1]Division of Bioinformatics and Biostatistics, National Center for Toxicological Research, US Food and Drug Administration, Jefferson, Arkansas

[2]Center of Pharmacology and Toxicology, Hannover Medical School, Hannover, Germany

---

6.1 INTRODUCTION

Drug development is a lengthy, complex, and costly process. While the number of new molecular entities approved for the past 50 years has remained around 20 to 30 per year [1], the total global spending on research and development (R&D) has doubled since 1996 to over $1.1 trillion [2] with an attrition rate of >99%. Only 0.1% of drug candidates that begin pre-clinical testing ever make it to human testing, and only one in five of these is ever approved for human usage [3]. There are several arguable causes for this trend, including increased regulatory barriers, the rising costs of scientific inquiry, and the lack of proven drug development models that have successfully and effectively incorporated new technologies such as genomics, bioinformatics, and cheminformatics. Above all, safety concerns continue to be one of the most significant causes for drug attrition in both the discovery and development processes [4–6].

It has been estimated that drug safety concerns alone account for approximately 35% to 40% of all drug discontinuations [7]. With the high drug attrition rate being due in large part to the high incidence of adverse effects, there is a huge financial incentive for drug companies to develop reliable approaches that allow rapid elimination of toxic compounds from the drug pipelines. Although common side effects of medications are detected and quantified in the premarket clinical trials, many safety issues cannot be detected until after a drug has

Y. Yao, B. Jallal, K. Ranade (Eds): Genomic Biomarkers for Pharmaceutical Development.
DOI: http://dx.doi.org/10.1016/B978-0-12-397336-8.00006-9

been on the market and used by a large and diverse population [8–10]. The recent advent of *in silico* (i.e., using computer software) methodologies and high-throughput *in vitro* (cell culture) screenings has alleviated some, but not all, of these challenges by providing a more efficient and effective way to develop drugs. The emerging research field technology known as toxicogenomics holds great potential for improving drug safety assessment, and consequently the drug discovery and development process. This chapter is aimed at discussing the potential applications and future challenges of toxicogenomics in drug discovery and development.

6.2 THE CONCEPT OF TOXICOGENOMICS

Toxicogenomics is a field of science that studies toxicology with genomics and other high-throughput molecular technologies and bioinformatics [11]. Typically, it applies transcript, protein and/or metabolite profiling technologies to investigate the interaction between genes/proteins/metabolites and environmental stress in disease causation and toxicity. The concept of toxicogenomics was first introduced in 1999, inspired by the rapid advancement of microarray technologies [12]. The field has quickly expanded by including proteomics, metabolomics and other new high-throughput technologies [13–15]. As a marriage of data-rich omics approaches with bioinformatics, toxicogenomics requires expertise from diverse fields ranging from toxicology, genetics, and molecular biology to computational science and bioinformatics [16–18]. The application of toxicogenomics offers an opportunity to identify the biological pathways and processes affected by exposure to pharmaceutical compounds and/or xenobiotics. It endeavors to elucidate molecular mechanisms involved in toxicity and to derive molecular expression patterns (i.e., molecular biomarkers) that predict toxicity or the genetic susceptibility [19–21]. Examining the patterns of altered molecular expressions caused by specific exposures can reveal how toxicants act and create their effects [22–24]. In addition, identification of toxicity pathways and potential modes of action allow for a more thorough understanding of safety issues. Increased understanding of a compound's mode of action can be used to predict toxicity. This ability is expected to reduce the attrition rate of new molecular entities and thus decrease the cost of developing new drugs. Indeed, there are high expectations that toxicogenomics in drug development will better predict/assess potential drug toxicity, and hence reduce attrition rates [25]. In the past 20 years, toxicogenomics has improved current approaches and led to novel predictive approaches for studying disease risk [26,27]. Most large pharmaceutical companies are now using the toxicogenomics approach as a predictive toxicology tool, with the goal of decreasing the drug-induced adverse reactions and thus reducing the drug attrition rate [28–30]. Toxicogenomics has enjoyed widespread attention as an alternative means of studying the underlying molecular mechanisms of toxicity and of addressing challenges that are difficult to overcome using conventional toxicology methods [31–33].

6.3 THE TECHNOLOGY LANDSCAPE OF TOXICOGENOMICS

Toxicogenomics utilizes several technology platforms with an emphasis on profiling transcripts (or genes), proteins, and metabolites (Table 6.1). Specifically, a wide range of biological assay platforms, such as transcriptomics and proteomics, have been employed in

TABLE 6.1   A Toxicogenomics Technology Landscape

Molecular Levels	Scientific Disciplines and their Representative Molecular Technologies	Data Type	Readout
RNA	Microarrays; next generation sequencing	Transcriptomics	Whole genome gene expression data
Protein	Resolution of complex mixtures for instance by 2D gel electrophoreses coupled with MS and MS/MS technologies and different ionization protocols	Proteomics	Proteome information
Metabolite	NMR and MS	Metabolomics	Metabolome information

constructing mechanistic and predictive toxicity. Furthermore, metabolomics uses endogenous metabolite signatures to determine the molecular mechanisms of drug actions and predict toxicity. Rapid advancement in mass spectrometry (MS) and nuclear magnetic resonance (NMR) has driven the development of proteomics and metabolomics, which complement transcriptomics [34–36]. However, in contrast to microarrays for transcriptomics, technologies for proteomics and metabolomics are limited by the huge range of analyte concentrations involved (at least six orders of magnitude), because most existing instrumentations are only capable of detecting more abundant species over those that are less abundant. This fundamental problem limits essentially all proteomic and metabolomic analyses to subsets of the complete collections of proteins and metabolites, respectively. Despite this limitation, proteomic and metabolomic approaches have fundamentally advanced the understanding of the mechanisms of toxicity and adaptation to stress and injury.

Two key advantages accrue when applying toxicogenomics. First, the large quantity and comprehensiveness of information that a single experiment can generate is much greater than that which traditional experiments generate. Second, the advancement of computing power and techniques enables these large amounts of information to be synthesized from different sources and experiments, and to be analyzed in novel ways. The sections below provide an overview of technologies used in toxicogenomics.

6.3.1 Transcriptomics

The transcriptome is the molecular compartment of transiently expressed genes. Transcriptomic technology is a means to detect the transcription of genetic information encoded within DNA into RNA (e.g., mRNA) molecules that are then translated into the proteins that perform most of the critical biological functions of cells [37]. Understanding the transcriptome is essential for interpreting the functional elements of the genome and revealing the molecular constituents of cells and tissues, and also for understanding development and disease. Transcriptomics can address all or a segment of the transcriptome, from normal or diseased single cells to tissues. Depending on the techniques employed, it can be used, for example, to catalog and annotate a cell's RNA complement, including coding as well as non-coding transcripts; to query the structure of the genes that give rise to them, including exon/intron boundaries, transcription start sites, splicing patterns and even gene fusion events; and to aid in mapping interactive networks. Perhaps most importantly, transcriptomics can be used to determine how the expression patterns of these transcripts change under different conditions (such as disease or drug treatment) and to develop biomarkers. The original emphasis on genome-wide mRNA transcript profiling is now extensively applied to other areas of study, such as evaluating different biological functions, understanding disease processes, the development of diagnostic markers, and therapeutic targets [38–40]. Further complexity in the biology of the transcriptome is aided by alternative mRNA splicing, the effects of epigenetics and epigenomics on the regulation of transcription, and posttranslational modifications. Various technologies have been developed to deduce and quantify the transcriptome. Generally speaking, they can be divided into two categories: hybridization-based and sequencing-based approaches.

6.3.1.1 *Microarray Technology – Hybridization-based Approaches*

Microarrays have been the stalwarts of high-throughput, genome-wide expression investigations since the mid-1990s [41,42]. This technology allows for rapid and simultaneous monitoring of tens of thousands of gene expressions in a single experiment. With the aid of sophisticated bioinformatics approaches, the amount of mRNA bound to the spots on the microarray is precisely measured, generating a profile of gene expression in the cell. These captured transcript changes reflect the activity of transcription across the whole genome at the defined time to a given condition. It allows hypothesis-generating research without any prior selection of candidate genes or gene sets.

Many commercial suppliers have made standard microarrays and analysis tools available [43,44]. The array platforms differ in design, manufacturing, hybridization, scanning, and data handling [45–47]. Nonetheless, the technology has notable limitations [48–51]:

1. It depends on existing knowledge about genome sequences and hence does not identify novel mRNAs – i.e., transcripts absent from databases cannot be detected.
2. It has a limited dynamic range, with the detection range limited by background for low expression transcripts and by saturation in the high expression transcripts.
3. Expression is indirectly measured as the scanned intensity of fluorescent hybridized probes, a process that may require complex normalizations.

To this end, specialized microarrays have also been developed to study different types of splicing, either to assess the effect of splicing events on transcript structure in a given sample or to compare the changes in transcript composition between two or more conditions (differential alternative splicing). There are three main array types available to study splicing, exon arrays [52,53], tiling arrays [54] and exon-junction arrays [55–58]. They are briefly described below:

- Exon arrays have probes designed to identify all the known or predicted exons that are expressed in a given cell or tissue sample [59,60]. They work similarly to gene expression microarrays except regarding the number and distribution of probes along all exons. These arrays have been used in recent years, not only to study splicing events but also as a substitute for gene expression arrays, given that they also allow the measurement of global expression using all probes [53,61].
- Genomic tiling arrays are designed to detect exon usage by setting contiguous matching probes at fixed distances across the genomes. These arrays, representing the genome at

high density, have been constructed and allow the mapping of transcribed regions to a very high resolution [62,63].

- Splicing junction microarrays are designed to measure connectivity of exons through the use of probes that span known exon junctions and can be used to detect and quantify distinct spliced isoforms. These arrays are usually custom-designed with the goal of identifying splicing events but require previous knowledge of splice junction positions [64–66].

As more transcriptional information becomes available, the array-based approach will become more complete and thus more informative. However, due to the limitations stated above and additional drawbacks, comparing expression levels across different experiments is often difficult and requires complicated normalization methods [67,68]. There is a general consensus that the ultimate approach to defining transcriptional complexity will come from the sequencing of full-length transcripts at quantitative throughput levels, which is becoming feasible due to next generation sequencing (NGS) technologies.

6.3.1.2 NGS Technologies – Sequencing-based Approaches for Transcriptomics Study

The arrival of deep sequencing applications for transcriptome analyses, RNA-Seq, may circumvent the above-mentioned disadvantages of microarray platforms. In contrast to microarray, transcriptome sequencing studies have evolved from determining the sequence of individual cDNA clones to more comprehensive attempts to construct cDNA sequencing libraries representing portions of the species transcriptome [69–72]. The use of sequencing technologies to study the transcriptome is termed RNA-Seq [73,74]. RNA-Seq uses recently developed deep sequencing technologies. In general, a population of RNA is converted to a library of cDNA fragments by use of adaptors attached to one or both ends. Each molecule, with or without amplification, is then sequenced in a high-throughput manner to obtain short sequences from one or both ends. In principle, any high-throughput sequencing technology can be used for RNA-Seq. This methodology has tremendously reduced the sequencing cost and experimental complexity, as well as improved transcript coverage, rendering sequencing-based transcriptome analysis more readily available and useful to individual laboratories. RNA-Seq technologies have demonstrated some distinct advantages over hybridization-based approaches such as microarrays that likely will enable them to dominate in the near future.

Currently, there are four major commercially available NGS technologies: Roche/454, Illumina HiSeq 2000, Applied Biosystems SOLiD, and Helicos HeliScope. Illumina's NGS platforms have a strong presence. Their sequencing-by-synthesis approach [75–78] utilizes fluorescently labeled reversible-terminator nucleotides on clonally amplified DNA templates immobilized to an acrylamide coating on the surface of a glass flow cell. The Illumina Genome Analyzer and the more recent HiSeq 2000 have been widely used for high-throughput massively parallel sequencing. In 2011, Illumina also released a lower throughput fast-turnaround instrument, the MiSeq, aimed at smaller laboratories and the clinical diagnostics market.

Although RNA-Seq is unlikely to completely supplant hybridization-based techniques in the near future, it offers a number of improvements over these technologies, for example:

1. unlike hybridization-based approaches, RNA-Seq does not depend on prior knowledge of the transcriptome, and is thus capable of new discovery and could reveal the precise boundaries of transcripts to single base precision [79];

2. the technique can also yield information about exon junctions, allowing the study of complex transcription units [80];
3. RNA-Seq has inherently low background and high sensitivity, and the upper detection limits are not constrained, together allowing the study of the transcription across a much wider range than for microarrays [56,81].

A discussion of the considerable differences between available RNA-Seq technologies is beyond the scope of this chapter. However these technologies share many common features. First, the RNA sample is either mRNA enriched or ribosomal RNA depleted. The choice depends on the intent of the experiment. A gene expression profiling experiment would enrich the mRNA and ignore the other RNA species, while an experiment focused on transcriptome characterization would deplete the ribosomal RNA leaving the mRNA, ncRNA, miRNA, and siRNA. Next, the RNA is fragmented and size selected. The size of RNA fragments required depends on the specific technology. Third, the fragments are reverse-transcribed into cDNA and are clonally amplified and tagged so that they can be attached to beads. The bead-bound fragments are then placed in a fluidics chamber, placed in the sequencer, and sequenced. The chemistry of sequencing varies between the platforms. However, each chemical change in the fluidics chamber (pH in the case of Ion Torrent, fluorescence for the other technologies) corresponds to a specific base and the sequence is recorded. The technologies described above all rely on the amplification of fragments via polymerase chain reaction (PCR), which will introduce bias and change the relative proportions of the RNA species present. Other technologies, referred to as 'single-molecule sequencing' or 'third-generation sequencing', avoid this amplification step and its attendant bias. However, these technologies have not yet been widely adopted by the scientific community.

Taking all of these advantages into account, RNA-Seq represents a paradigm shift in transcriptomics studies, with concomitant benefits for toxicogenomics. This technology has already been extensively applied to biological research, resulting in significant and remarkable insights into the molecular biology of cells [82–84]. The pharmaceutical industry has already embraced sequence-based technologies, and it is likely that these technologies will have their impact throughout the drug discovery process [85–87].

6.3.2 Proteomics and Metabolomics

Proteomics is a methodology that attempts to compare and quantify changes of protein profiling in a system-wide proteome analysis to evaluate the overall cellular response to drug treatments [88,89]. Several proteomic technologies are widely used and these techniques are complementary since they focus on subsets of proteins that are only partially overlapping. The main distinction of these technologies resides in their being either gel-based or gel-free techniques using liquid chromatography tandem mass spectrometry (LC-MS/MS) [90,91]. The possibilities of proteomics and its promising results for improving current predictive and mechanistic toxicological studies have been reviewed elsewhere [36] and [92–94].

The majority of small molecule drugs and biologics act on protein targets. These proteins do not act in isolation, but are embedded in cellular pathways and networks and are tightly interconnected both physically and functionally with many other proteins and cellular components [95]. Thus, pre-clinical stages in the drug discovery process require a multitude of biochemical and genetic assays in order to characterize the effects of drug candidates on cellular systems and model organisms. Given this complexity, it is natural to apply proteomics in the drug discovery process to understand the effects of drug candidates on their protein targets and shed light on the cellular mechanisms resulting in the observed phenotype. Over the last 15 years, proteomics technology has made significant progress in several areas and has become an important tool at various stages in drug discovery [96].

The terms metabolomics and metabonomics are often used interchangeably [34]. Metabolomics technology has been part of drug discovery and development for over a decade, and has gained early and wide acceptance as a novel tool for the rapid elucidation of mechanisms and biomarkers in pre-clinical discovery [97,98]. It allows comprehensive and simultaneous profiling of hundreds of discrete biologically important molecules, including amino acids, sugars, lipids, and exogenous substances from biological fluids and tissues [99,100]. Metabolomics is one of the 'omics' approaches that mostly represents the interplay of internal biological regulation and external environmental influences on disease, thereby being of particular importance to disease mitigation and management [101].

Since many metabolites are species-independent and evaluated in non-invasively obtained biofluids, metabolomics-derived biomarkers can potentially have a large impact on clinical and translational science [102]. Early on, translational biomarkers for drug safety demonstrated the promise of metabolomics approaches [103–105]. It was reported that pre-dose metabolomic measures in rat urine could be used to predict the postdate outcome to acetaminophen treatment [106]. Further research demonstrated a similar phenomenon in humans where samples taken shortly after dosing with acetaminophen predicted later hepatic sensitivity [107]. Under a recent Food and Drug Administration (FDA) Guidance for Industry, biomarkers discovered using metabolomics may be submitted for consideration as new drug development tools and entered into a formal biomarker qualification process [108].

6.4 TOXICOGENOMICS IN DRUG DISCOVERY AND DEVELOPMENT

Drug development has become much more difficult and challenging than ever, as both scientific advances and regulatory concerns have led this process to become more rigid, more expensive, and more risky [109]. Drug toxicity and adverse drug reactions (ADRs) have been major problems in the development and clinical applications of drugs. Currently, there is no simple way to predict whether a patient will respond well, badly or have no response to a drug. There is no 'one size fits all' system for the pharmaceutical companies in developing drugs [110–112]. To address this issue, toxicogenomic studies have been applied early in the drug discovery pipeline for the last two decades, often using a classification approach to group novel and reference compounds so as to predict potentially the toxicity of drug candidates [113,114]; the industry is now target and hypothesis rich. The central challenge for today and for the future will be to identify chemicals of sufficient specificity by exploiting modern molecular biological insights.

6.4.1 The Roles of Toxicogenomics in Drug Discovery and Development

Gene expression analysis applied to toxicology studies has often been used by the pharmaceutical industry as a useful tool to identify safer drugs in a quicker, more cost-effective manner. Monitoring gene expression through the dual approaches of transcriptomics (RNA profiling) and proteomics (protein profiling) has become a key component in our efforts to understand complex biological processes. From the molecular stratification of disease states and the selection of potential drug targets, to patient selection and the confirmation of engagement of pharmacology in clinical studies, we are seeing the impact of gene expression profiling across all phases of the drug discovery and development process [112,115,116]. Studies have already demonstrated the benefits of applying gene expression profiling towards drug safety evaluation, both for identifying the mechanisms underlying toxicity, and for providing a means to identify safety liabilities early in the drug discovery process [117,118]. Furthermore, toxicogenomics has the potential to better identify and assess ADRs of new drug candidates or marketed products in humans [11]. As depicted in Fig. 6.1, toxicogenomics can greatly influence the drug development process by:

1. increasing our knowledge of molecular mechanisms of toxicity and efficacy;
2. identifying sensitive biomarkers for better monitoring compound toxicity in clinical trials;
3. enabling more informed decisions regarding safety of compounds;
4. enhancing the ability to extrapolate accurately between experimental animals and humans in the context of risk assessment; and
5. providing a better understanding of the influence of genetic variation on toxicological outcomes.

6.4.2 Toxicogenomics Applications – A Liver-Dominated Field

The liver is one of the major organs for the synthesis and secretion of substances which metabolize endogenous and exogenous materials. There has been a great deal of interest in elucidating the predictive and mechanistic genomic biomarkers of hepatotoxicity [119–121]. Serum alanine aminotransferase (ALT) or aspartate aminotransferase (AST) measurements are typically used to monitor liver damage and classify samples into responders and nonresponders upon an exposure to a toxicant [122]. However, increases in the transaminase are not good prognosticators of liver injury and as such, have limitations in their usage as biomarkers. For instance, ALT measurements do not always correlate well with histopathological data. There are cases where the variation of the ALT measurements among samples sharing the same necrosis severity score is large [123]. In a more practical approach, the severity of liver injury is represented by a composite score that incorporates different measurements or according to the similarity of the biological processes of the samples to reflect a phenotypic representation of a toxicant effect [123,124]. To improve the predictability of drug-induced liver injury (DILI), it is essential to combine the conventional approaches with the emerging technologies of toxicogenomics. Currently, there does not exist any reliable strategy for preventing DILI [125]; treatment options are limited to discontinuing the offending drug, supportive care, and transplantation for end-stage liver failure [126]. Thus, it is crucial to develop methods that will detect potential hepatotoxicity among drug candidates as early and as quickly as possible. A better prediction, characterization, and

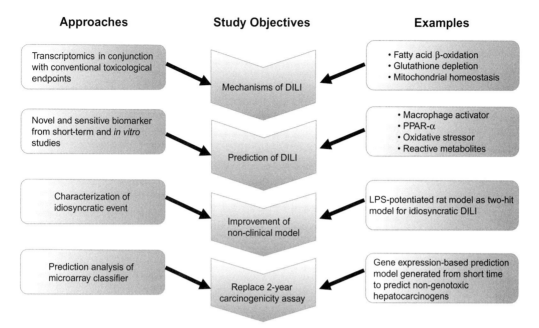

FIGURE 6.1 Toxicogenomics applications exemplified by study of drug-induced liver injury (DILI) and liver carcinogenicity in drug discovery and development. (1) Microarray analysis helps elucidate mechanisms of DILI with gene expression profiling coupled with conventional toxicology endpoints. For example: gene expression analysis has been reported to reveal the mechanisms of hepatic steatosis. (2) Toxicogenomics represents an approach to develop novel biomarkers that can be used as sensitive indicators of DILI. For example, macrophage activator, PPAR-α, oxidative stressor (OS) or reactive metabolites (RM), all produce oxidative stress and liver injury in rat, but through different mechanisms and resulting in different toxicological outcomes. (3) Genomics tools have been used to investigate the mechanisms underlying various idiosyncratic DILI. For example, toxicogenomic approaches have been used to evaluate the LPS-potentiated rat model. This model is a two-hit model for idiosyncratic toxicity. It shows that genomics analysis is a valuable method for hypothesizing mechanisms and discovering specific transcript changes that can be exploited as biomarkers for idiosyncratic DILI. (4) Prediction analysis of microarray classifier generated from short-term animal study can be used to predict potential non-genotoxic hepatocarcinogenicity. It was thought that a mechanism-based strategy should be employed in order to obtain useful biomarker genes for hepatocarcinogenicity. This shed light on the possibility of replacing two-year carcinogenicity study with toxicogenomics.

understanding of DILI could result in safer drugs and significantly reduce the cost of drug development. However, the complexity of the hepatotoxicity endpoint makes it very difficult to be predicted with the current approaches for pre-clinical assessment.

While toxicogenomics has been applied in the diverse areas of toxicology, the majority of applications (or data) so far have focused on the study of hepatotoxicity. Along with the reasons stated above, liver toxicity alone accounts for 40% of drug failures in clinical trials and 27% of market withdrawals [127–129]. DILI has become the most common cause of fulminant hepatic failure in patients, both in the United States and Europe [130–133]. It accounts for approximately one half of acute liver failure and transplantation in Western countries [134,135]. DILI also has frequently been the single reason for denial of new drug approvals by the United States FDA [136–138]. Failures due to liver toxicity span the pre-clinical and clinical development phases, and unfortunately have been an insidious cause of withdrawals post-market. Hepatic toxicology studies that have employed toxicogenomic technologies to date have already provided a proof-of-principle for the value of toxicogenomics in drug discovery and development [23] and [139–141]. Notably, in most cases, gene expression methodologies identified the differentially regulated genes that gave useful insight into the mechanism of toxicity [142–144]. Beyond this application for interrogating mechanisms of hepatotoxicity, toxicogenomics represent a viable approach for developing novel biomarkers that can be used as sensitive indicators of DILI. Hepatic toxicity is traditionally identified through the integrated evaluation of histopathological findings and clinical pathology parameters, such as ALT and/or AST [145]. However, these changes often occur only after prolonged periods of dosing, and may be too subtle to interpret in early, short-term studies [146]. In contrast, genomic biomarkers are frequently more sensitive than the traditional functional and morphological markers [147,148], providing the prospect of earlier detection of hepatotoxicity in short-term studies, prior to full phenotypic manifestation.

As an example, acetaminophen (APAP) is a well-known hepatotoxicant and the most common cause of acute liver toxicity in Europe and the US [135] and [149]. The traditional biomarkers such as ALT and AST are positive after pronounced liver injury has occurred. It is well worth noting that diagnostic gene expression change in rat blood cells following APAP exposure occurs well before liver damage can be diagnosed by classical parameters [150]. Further study in humans with a low dose of APAP demonstrated that blood transcriptomics are sufficiently sensitive and robust for prevention of liver injury, far outperforming classic clinical chemistry tests [23]. The transcriptome may be a more effective marker or predictor for DILI than other biomarkers. The use of transcript profiling for distinguishing classes of toxicity has become well established [34] and [151,152]. Several toxicogenomics studies have monitored the gene expression changes in the liver and/or the blood following exposure to hepatotoxicants.

6.4.3 Predictive Toxicology with Toxicogenomics

An early and reliable prediction of a drug candidate with potential to induce toxicity represents one of the major challenges in drug development. Toxicogenomics can inform the drug development process by predicting an organism's response to a toxicant with genomics. Predictive toxicogenomic studies usually compare the gene expression changes caused by chemicals with unknown toxic potential to the profiles for model compounds with known toxicity. To use toxicogenomics as a predictive tool, the prior knowledge of gene expression patterns related to toxicity is absolutely necessary. Currently, routine pre-clinical safety assessment relies on a specific set of parameters consisting mainly of serum biochemistry, hematology, and histopathology [153]. Unfortunately, some of these parameters are neither sensitive nor specific enough for certain organ toxicities [154,155]. A means to overcome this limitation is the integration of molecular profiling-based toxicogenomics technologies into regular toxicological assessments. Specifically, global gene transcriptional profiling has the potential to predict toxic responses under the principle that compounds that induce toxicity through similar mechanisms will cause similar changes in gene expression patterns [156]. By grouping the gene expression profiles of well-characterized model compounds and phenotypically anchoring these changes to conventional indices of toxicity, a gene expression signature or fingerprint related to specific organ toxicity could be generated and used to predict the toxicity of a candidate drug. In this manner, gene expression profiles might be seen as unique drug 'fingerprints', the idea being that if a fingerprint closely matches that of a similar toxicant in a database, there is a suggestion of toxicity. This approach may provide a novel metric for the identification and ranking of the potential toxicity of candidate drugs, thus increasing the likelihood of success in pre-clinical and clinical development [157]. The predictive capacity of gene expression profiling has been demonstrated in several studies [158–160]. In fact, some pharmaceutical companies have built their own databases to support drug development. If successful, this approach would improve the identification of potentially toxic compounds and thus increase the likelihood of success for those candidate drugs passing the screening. Hepato- and nephrotoxicity are two related attrition factors in pre-clinical drug development and thus a main focus of study with toxicogenomics [29] and [161–163]. Specific gene expression profiles identified to be associated with various hepato-toxicants have been reported, including acetaminophen, carbon tetrachloride, and TCDD and more [164–166]. Furthermore, these gene expression signatures may be linked to specific biological processes, such as macrophage activators, peroxisome proliferators, and oxidative stressors/reactive metabolites [167,168]. Studies have also demonstrated the capability of toxicogenomics in the prediction of nephrotoxicity and identification of candidate toxicity-related biomarkers in rat kidney [169,170].

Toxicogenomics can also be used with gene expression data from a short-term *in vivo* study or *in vitro* study to predict outcomes in long duration studies, such as chronic or carcinogenicity toxicity studies [171,172]. Preliminary toxicogenomic studies suggest that gene expression patterns are often predictive of phenotypic alterations [173]. The carcinogenic potential of chemicals is currently evaluated with rodent life-time bioassays, which are time consuming, and expensive with respect to cost, number of animals, and amount of compound required. Using gene expression profiles from the livers of rats treated for up to 14 days with hepato-carcinogens, gene expression patterns were identified and developed into classifiers to predict a set of independent validation compounds; the prediction accuracy is reported to be up to 88% [172]. In addition, gene expression patterns were found to be more sensitive for detecting renal toxicity than the biochemical and histopathological parameters used in conventional toxicology testing [18] and [174]. Predicting the long-term effects of compounds using short-term assays is a feasible approach [140]. Therefore, it is believed that toxicogenomics could accelerate the process of drug discovery and development [175].

6.4.4 Mechanistic Toxicology with Toxicogenomics

The advent of genomic technologies has facilitated major advances in our understanding of the molecular details of biology and holds the promise of providing new insights into elucidating the molecular mechanisms of a variety of toxicities. The application of toxicogenomics to mechanistic studies has played an important role when the toxicity of candidate drugs is not associated with well-established biomarkers or significant morphological changes [176,177]. Modern mechanistic analysis leverages the increasing sophistication of functional genomics to interpret the biological functions of particular gene sets, understand how the gene function contributes to a biological process, gain information on likely initiating and controlling factors, and predict possible biological consequences of disturbed expression patterns [141]. Mechanistic toxicity has application to a wide range of drug safety issues, and offers an opportunity for improving the understanding of biological pathways that are modulated by a drug. Such a need is apparent when it comes to achieving a better assessment of safety issues throughout the entire drug development cycle [178–180]. Gene expression profiling identifies critical, toxicologically-relevant genes and signal-response pathways, and promises to improve risk assessment and safety evaluation practices. By examining alterations in gene expression in response to drugs, it is possible to generate hypotheses about the underlying mechanisms of toxicity, which could be crucial for the identification of potential safety liabilities early in the drug development process [181–183]. Public annotations of functional and biological aspects of the rat and human genome such as cellular, molecular, and biological components have increased the level of understanding of complicated co-regulation within and among pathways and genes [184,185]. When transcriptional data was linked to phenotype, toxicogenomics became very useful in predicting drug-induced toxicity and understanding the underlying mechanisms.

A great deal of mechanistic toxicogenomics research has focused on hepatotoxicity because a wealth of published and proprietary gene expression information is available for the liver. Clearly, the lessons learned from the study of hepatotoxicity may be extensible to other tissue and organ toxicities, such as those for the kidney, heart, and bone marrow [186–189]. About a decade ago, by using microarrays in rat hepatocytes, Waring et al. reported that the mechanism of action for prototypical toxicants could be discerned [190]. With the assumption that drug toxicity is accompanied by transcriptional changes in gene expression that are causally linked to downstream of the toxicity [191], Waring and his colleagues conducted a large-scale gene expression analysis in order to gain an enhanced understanding of the pathogenesis of drug-induced phosphilipidosis using the human hepatoma HepG2 cell line. This study established four affected pathways which contribute to the formation of phosphilipidosis [192]. Altered gene expression was consistent with lysosomal phospholipase inhibition with increased expression of phospholipid degradation genes such as N-acylsphingosine amidohydrolase 1 and sphingomyelin phosphodiesterase. There was also a role for reduced lysosomal enzyme transport demonstrated by decreased expression of genes such as adaptor-related protein complex 1 sigma 1 that transports lysosomal enzymes between the Golgi network and the lysosyme. The results also indicated a role for increased phospholipid and cholesterol biosynthesis (stearoyl-CoA desaturase and HMGCoA synthesis, respectively), both of which are triggers for phosphilipidosis [193]. Thus, phosphilipidosis results from the combination of events involving both increased synthesis and decreased degradation of phospholipids. In addition, the extension of this study identified a set of 12 marker genes for predicting phosphilipidosis [194].

In another example, Burczynski et al. used HepG2 to profile cellular gene expression after human exposure to mechanistically unrelated drugs (cytotoxic anti-inflammatory drugs and DNA-damaging agents) with a low-density DNA microarray containing a set of 250 human genes [195]. The study was able to identify a reproducible, small, common set of genes from two classes of drugs. This gene set was used to distinguish compounds from these two classes based on a cluster analysis [195]. Discriminatory analysis identified genes involved in DNA repair, xenobiotic metabolism, cell-cycle control, apoptosis and transcriptional activation. Together with others, these early transcription-profiling studies demonstrated that drug treatment can trigger expressional alterations in many genes which may be linked to putative modes of action associated with adverse events [196].

Mechanistic toxicogenomics also showed advantages over conventional toxicity measures in a study of cardiotoxicity [197]. During a two week study at a high dose of a cardiotoxicant, rats showed myocardial degeneration and necrosis. At lower and middle doses, however, there was no evidence of cardiotoxicity using the traditional toxicological end points [198]. A mechanistic toxicogenomic investigation revealed only minor gene expression pattern alterations in the heart from the low dose group. However, rats given high dose for one and five days showed striking, similar gene expression alterations, even though there were no clinical signs or symptoms in rats treated with the high dose on day one. Thus, the alteration in gene expression provided earlier detection than the traditional toxicologic endpoints. To explore these results further, the researchers were able to determine that a number of the differentially regulated genes were related to mitochondrial impairment. The authors emphasized that the gene expression assay can be used to generate a hypothesis – that the mechanism of toxicity was the inhibition of mitochondrial function. Further studies are needed to confirm this hypothesis.

6.5 EFFORTS TO ADDRESS TECHNICAL CHALLENGES IN TOXICOGENOMICS

There are many technical issues that need to be addressed when using high-throughput technologies such as transcriptomics, proteomics, and metabolomics for toxicogenomics. For example, microarray experiments are fraught with potential sources of variation including multiple options of the process that alter the target quality [199,200]. In particular, the comparability and reliability of microarray gene expression data across laboratories has previously been questioned [201–203]. The overall quality of a specific array design also depends on the consistency of manufacturing and the limits of the platform's dynamic range [204–207]. To address these concerns, the microarray community and regulatory agencies have developed a consortium to establish a set of quality assurance and quality control criteria to assess and assure data quality, to identify critical factors affecting data quality and to optimize and standardize microarray procedures so that biological interpretation and regulatory decision making are not based on unreliable data. These fundamental issues were addressed by the MicroArray Quality Control (MAQC) project consortium (http://edkb.fda/gov/MAQC/, accessed on Nov. 18, 2012). The MAQC project originally aimed to establish quality control metrics and thresholds for the objective assessment of the performance achievable by different microarray platforms [208]. Meanwhile, the project addressed the parallel issues related to genome-wide association studies (GWAS), biomarker development and outcome prediction and, more recently, NGS. In addition to addressing data quality metrics, the MAQC project evaluated the merits and limitations of various data analysis methodologies [209]. It confirmed that with careful experimental design and appropriate data transformation and analysis, microarray data are reproducible and comparable across different platforms from different laboratories. It is anticipated that the MAQC projects will help to improve microarray and other emerging biomarker technologies and foster their appropriate use in the discovery, development and review of FDA-regulated products. The results of these efforts are published in a compendium of papers [210–212]. At the time of this writing, the MAQC consortium had completed the third phase of MAQC, also known as Sequencing Quality Control (SEQC). The project goals and practices of this project parallel those of the previous two MAQC projects (i.e., MAQC-I and -II) with the overall aim of establishing a set of recommendations for applying NGS technologies in the areas of clinical research, patient care and safety evaluations including toxicogenomics.

6.6 PUBLICLY AVAILABLE TOXICOGENOMIC DATABASES

With the wealth of the genomic data generated from many microarray experiments, investigators quickly realized that databases and analytical tools were essential to effectively manage and condense the data into a more manageable form. In addition to the challenges in the data analysis and interpretation of large databases, there is a consensus among the scientific community for the need of a predictive toxicogenomic database [213–215]. Building on the momentum gained from leveraging databases and computational algorithms for genome sequencing efforts, engineers, statisticians, mathematicians, and computer scientists have developed analytical tools and shared resources for microarray gene expression data. These data warehouses provide a means for the scientific community to publish and share data from large-scale experiments in order to advance understanding of biological systems. The data repositories would also serve as a resource for data mining and discovery of expression patterns common to certain experimental conditions, phenotypes and diseases. Such repositories could also serve the regulatory community as a body of knowledge that could be compared with toxicogenomics data submitted as part of the compound registration process. Importantly, some public repositories consist in the promotion of international standards in data organization and nomenclature.

Although several reports have described software for managing genomics/transcript profiling data at the local or laboratory level, there are compelling reasons for the establishment of public databases that house not only such transcript profiling data but also the corresponding classical toxicological endpoints. The utility of such a public toxicogenomics data repository largely depends on the proper functional structuring of the data in a relational schema that allows efficient extraction of relevant information from it. Data should undergo rigorous curating, and provenance and experimental metadata should be duly incorporated. Ideally, tools will be integrated for quality assurance (QA), annotation, flexible query, graphical display, and a broad array of statistical, pattern recognition, and machine learning analytics for interpretation and model construction. Generally, the value of the repository will expand enormously as the types and numbers of drugs and chemicals as well as the number of associated data types and endpoints increase. More valuable still will be gene expression data collected alongside corresponding conventional toxicological endpoints such as organ weight, clinical chemistry, hematology and histopathology. A high quality database and robust software with appropriate algorithms for the comparison of complex gene expression fingerprints are vital for the interpretation and utilization of the toxicogenomic data. By combining conventional toxicology phenotypes, validated signatures can then be used for predictive and mechanistic toxicity studies.

Several toxicogenomic databases are currently being built [216–218] and will be briefly reviewed in this chapter. Summaries of the corresponding experiments designs are given in Table 6.2.

TABLE 6.2 Summary of Three Databases (DrugMatrix, TGP and PredTox)

	DrugMatrix (Iconix) (USA)		TGP (Japan)			Human donor	PredTox (EU)
Species	Male Sprague-Dawley Rat		Male Sprague-Dawley Rat			Human donor	Male Wistar rats
Study type	*in vivo*	*in vitro*	*in vivo*	*in vitro*	*in vitro*	*in vitro*	*in vivo*
Dose type	Daily repeated	Single	Daily repeated	Single	Single	Single	Daily repeated
Dose level	2 dose levels (maximum tolerated dose/fully effective dose)	2 dose levels (low, high)	3 dose levels (low, middle, high)				2 dose levels (low, high)
Duration	1, 3, 5 days plus some 7 days	6, 24 h	3, 7, 14, 28 days	3, 6, 9, 24 hr	2, 8, 24 hr	2, 8, 24 hr	1, 3, 14 days
# tested compounds in liver	343 chemicals (mostly drugs)	126 chemicals (mostly drugs)	131 chemicals (mostly drugs)			119	14 proprietary drug candidates and two reference toxic compounds
Other tissues	Kidney, heart, thigh muscle, bone marrow, spleen, brain, intestine	NA	Kidney				Kidney, blood, urine
Microarray Platform	Affymetrix RG230–2.0 array and CodeLink RU1 array		Affymetrix RG230–2.0 array			Affymetrix human U133 plus 2.0	Affymetrix RG230–2.0 array
# arrays	CodeLink: 5149 arrays Affymetrix: 2216 arrays	~800	6264	6249	3140	2004	2300
Clinical information	Histopathology; body/organ weight; food consumption; hematology and blood chemistry	NA	Histopathology; body/organ weight; food consumption; hematology and blood chemistry			NA	Histopathology; body/organ weight; food consumption; hematology and blood chemistry

Iconix Pharmaceutical (now a part of Entelos, San Mateo, CA) teamed with its partners and built a database tracking rat gene expression profiles associated with toxic response based on short-term, repeated-dose studies [219]. The database encompasses marketed and withdrawn drugs, toxicants, and reference standards. The effort was aimed at providing requisite data to improve pre-clinical drug safety and toxicity assessment in order to reduce the late stage attrition rate in drug discovery and development. DrugMatrix was initially constructed using the CodeLink RU1 rat microarray platform (GE Healthcare, Chalfont St Giles, UK), and then extended to study transcriptional effects on the whole genome RG230_2.0 rat GeneChip array platform (Affymetrix, CA, USA). The most recent release of DrugMatrix, now owned and publicly released by the US National Institute of Environmental Health Sciences (https://ntp.niehs.nih.gov/drugmatrix/index.html, accessed on Nov. 22, 2012), contains data on some 636 compounds. Gene expression profiling was conducted for seven different tissue types (liver, kidney, heart, bone marrow, thigh muscle, spleen, and intestine), and for rat primary hepatocytes, for a total of more than 4100 drug-dose-time-tissue combinations. The extensive database allows for improved mechanistic understanding and interpretation of drug signature through the availability of more curated genes within the whole genome. The analysis of array data showed that the compounds with low margins in a four-day rat toxicology study were predicted to have phospholipase D (PLD) and hepatotoxicity [220]. PLD is an enzyme that catalyzes the hydrolysis of phosphatidyl choline (PC) to generate phosphatidic acid and choline. GPI-PLD [glycosylphosphatidylinositol (GPI)-specific phospholipase D (PLD)] is a secreted mammalian enzyme that specifically cleaves GPI-anchored proteins. In addition, the enzyme has been shown to cleave GPI anchor intermediates in cell lysates [221,222]. The expression of PLD may be associated with PPARα-induced hepatotoxicity through a complex interaction with nuclear receptors including CAR and PXR [222,223]. DrugMatrix has been applied to assist scientists in the selection of the leads and drug candidates at the earliest and most-cost-effective stages of drug discovery. Recent publications have demonstrated that coupling gene expression profiling with traditional toxicity measurements provides a powerful tool for understanding individual compound effects in rat.

The Japanese Toxicogenomics Project is a five-year collaborative project (2002–2007) initiated by a consortium consisting of the Japanese government and 15 pharmaceutical companies. The project produced the 'Toxicogenomics Project-Genomics Assisted Toxicity Evaluation system' (TG-GATEs), a large-scale database of transcriptomics and pathology data potentially useful for predicting the toxicity of new chemical entities [224]. Conventional *in vivo* toxicology data were collected from both single and repeat dosing studies on rats, and the gene expression was measured for the liver (and kidney in some cases). To provide information on species differences, gene expression was also measured *in vitro* in rat and in human hepatocytes treated with the chemicals. Approximately 130 chemicals, primarily medicinal compounds, were tested at multiple doses. Gene expression was analyzed using Affymetrix GeneChip arrays (Hgu133plus2 and rat233a). For each compound, the gene expression data were collected within four separate experiments: *in vitro* human hepatocytes exposures, *in vitro* rat primary hepatocytes exposures, liver and kidney from *in vivo* single dose rat exposures, and liver and kidney from *in vivo* repeat dose rat exposures. Within *in vitro* exposures, duplicate samples were collected at four doses (control, low, medium, high) across three time points (2, 8, and 24 hour exposure). Within the single dose *in vivo* exposures, triplicate samples were collected at the four doses across four time points (3, 6, 9, and 24 hours). Within the repeat dose *in vivo* exposures, samples were collected at 4, 8, 15, and 29 days following daily dosing with the same four doses. These data were initially made available through the TG-GATE website (http://toxico.nibio.go.jp/open-tggates/search.html, accessed on Dec. 2, 2012). They have also been made available through ArrayExpress in the form of bulk downloads of array data from each experiment type: *in vivo* single dose liver and kidney (E-MTAB-799), *in vivo* repeat dosing liver and kidney (E-MTAB-800), *in vitro* rat hepatocytes (E-MTAB-797), and *in vitro* human hepatocytes (E-MTAB-798). This citation also provides a description of the Japanese Toxicogenomics Project.

Meanwhile, one of the integrated EU Framework 6 Projects entitled Predictive Toxicology (PreTox) tackled toxicogenomics using a systems toxicology approach. The project was coordinated by the European Federation of Pharmaceutical Industries and Associations (EFPIA), a body representing the research-based pharmaceutical industries and biotech Small and medium-sized enterprises (SMEs operating in Europe. It is partly funded by the European Commission Life Sciences, Genomics and Biotechnology for Health Priority with a grant of €8M aimed at assessing the value of combining results from 'omics' technologies (genomics, proteomics and metabolomics) together with the results from conventional toxicology methods, for more informed decision making in pre-clinical safety evaluation. The effects of 16 test compounds were characterized using conventional toxicological parameters and 'omics' technologies. The three major observed toxicities, liver hypertrophy, bile duct necrosis and/or cholestasis, and kidney proximal tubular damage, were analyzed in detail [225,226]. The combined approach of 'omics' and conventional toxicology has proved to be a useful tool for mechanistic investigations and the identification of putative biomarkers [29].

6.7 BIOINFORMATICS CHALLENGES IN TOXICOGENOMICS

The exponential growth in the amount of toxicogenomics data being generated poses both challenges and opportunities, i.e., data management and extraction of useful information from these data, and the development of tools and methods capable of transforming these data into knowledge. Both the generation and validation of hypotheses require not only a comprehensive description of the different components of an experiment (such as cell/organ exposure, sample collection, and processing components of the experiments) in the database, but also complex computational and bioinformatics approaches. Bioinformatics is an interdisciplinary field that develops and improves upon methods for storing, retrieving, organizing, and analyzing biological data [227]. It is a bridge between observation (experimental data) in diverse disciplines of biology and genomes and extrapolation of information, by computational means, about how the systems and processes function [228]. One objective of bioinformatics analysis is to extract useful knowledge from the flood of data, including biological texts, for the purpose of further analysis leading ultimately to useable knowledge [36] and [229–231].

Due to the nature and characteristics of the diverse techniques that are applied for biological data acquisition, and depending on the specificity of the domain, biological data might require a number of preparatory steps prior to analysis. These are usually related to the selection and cleaning, preprocessing, and transformation of the original data. The data preprocessing task is subdivided into a set of relevant steps that could improve the quality and success when applying, for example, machine learning techniques [232,233]. They refine/depurate the data to make certain data mining operations more tractable. There are three well-known data preprocessing topics that are among the most applied. These are missing value imputation, data normalization, and discretization [234,235].

Once data have been prepared for analysis, the data mining discipline offers a range of techniques and algorithms for the automatic recognition of patterns in data. Depending on the goals of the study and the nature of the data, these techniques have to be applied differently. Data mining techniques provide a robust means of evaluating the generalization power of extracted patterns in unseen data, although these must be further validated and interpreted by the domain expert [236,237]. Machine learning is the most representative task of many data mining applications [237,238]. It is essentially a set of computer programs that make use of sampled data or past experience information to provide solutions to a given problem. The most broadly applied machine learning types are supervised learning and unsupervised learning [239–241]. Supervised classification, also known as class prediction, is a key topic in the machine learning discipline. Its goal is to construct a function (or model) to accurately predict the target output of future cases whose output value is unknown. Supervised classification starts with a set of training data that consists of pair of input cases and desired outputs to derive a predictive model and the model then can be subsequently used to predict the outcome of unknown samples [242]. Supervised classification techniques have been shown capable of obtaining satisfactory results in toxicogenomics [243–245]. On the other hand, unsupervised learning normally involves clustering, that is, the partitioning of samples into subsets (clusters) so that the data in each cluster shows a high level of proximity [246]. A wide range of machine learning methods have been proposed by the data mining community in recent decades. Excellent reviews in the literature are available outlining their use in high-throughput genomic data analysis [101] and [247–251].

The bioinformatics community has 'borrowed', customized and developed a large number of applications and resources during the last few years in order to deal with the data produced by the high-throughput technologies. These include complex statistical methods and integrated analysis approaches for identifying and classifying patterns of gene expression changes, and some of these have been extensively employed for correlating gene expression profiles to toxicity [252–255]. Although it is still difficult to find singular off-the-shelf software that meets all the needs for toxicogenomics, many free and commercial software and bioinformatics tools serve in a superior manner to discover biologically meaningful knowledge. Readers can easily find a rich collection of resources showing how to conduct microarray data analysis [256–258]. As drug development companies are incorporating high-throughput technologies for safety assessment, so too the FDA is developing its own tools for review purposes. For example, the FDA has developed ArrayTrack™, a comprehensive microarray data management, analysis and interpretation system (http://www.fda.gov/ScienceResearch/BioinformaticsTools/Arraytrack/default.htm. Last accessed on Feb. 25, 2013) to support the FDA's Voluntary Genomics Data Submission (VGDS) program.

6.8 TOXICOGENOMICS IN REGULATORY APPLICATION

By recognizing the impact of genomics and other 'omics' technologies on drug development and eventually on the regulatory process, the FDA coined the concept of 'safe harbor' in early 2000, to describe a novel way to share information between the FDA and external scientists (e.g., industry scientists, academic researchers) [259]. The concept evolved into a VGDS mechanism and is defined in the FDA's Guidance for Industry: Pharmacogenomics Data Submission (http://www.fda.gov/downloads/RegulatoryInformation/Guidances/ucm126957.pdf, accessed on Jan. 8, 2013). The guidance itself encourages the integration of biomarkers in drug development and their appropriate use in clinical practice. It is believed that this approach will help to alleviate stagnation and to foster innovation in the development of new medical products, and, ultimately, lead to better medicine. Furthermore, the guidance offers the FDA's current view on pharmacogenomics and what the regulatory agency believes are the scientific grounds for evaluating such information as it relates to voluntary versus required submission of data. Many principles found in this guidance apply to toxicogenomic studies. In particular, the identification, evaluation, and validation of biomarkers are critical components of every pharmacogenomic and toxicogenomic case study in regulatory decision making. The guidance is general and encompasses both genetic and genomic biomarkers; e.g., a CYP2D6 (cytochrome P450 2D6) mutation and an increase in HER2 (human epidermal growth factor receptor 2) expression can be viewed as a genetic and a genomic biomarker, respectively. The VGDS program emphasized by the guidance creates a forum for scientific data exchanges and discussions with the FDA outside of the regular review process. This is necessary, since future data submissions will contain many more complex gene expression profiles and large-scale single nucleotide polymorphism maps (e.g., from whole genome scans), which will present new challenges in defining the analytical and clinical validity of such new and highly complex biomarker sets. Based on the knowledge on genomics data related to drug safety, the FDA has approved 117 drugs so far with pharmacogenomics information in their label (http://www.fda.gov/drugs/scienceresearch/researchareas/pharmacogenetics/ucm083378.htm, last accessed on Feb. 18. 2013).

Although good progress has been made in recent years, additional proof-of-principle studies are needed for the regulatory community to become more acceptant of using toxicogenomic data as part of the regulatory decision-making process. It will be important to demonstrate, for instance, that toxicogenomics not only can confirm what is already known about specific compounds and toxic end points (i.e., phenotypic anchoring), but also can accurately predict toxicity for unknown compounds.

6.9 THE FUTURE PERSPECTIVE OF TOXICOGENOMICS

The excitement that surrounded toxicogenomics when it was a new field has waned in recent years [33]. While its objectives and goals have so far remained unchanged, a shift in emphasis can be anticipated in the future. It is our view that toxicogenomics will remain a valuable tool for mechanistic studies. However, toxicogenomics will need to fully take advantage of data richness from enabling technologies with improved tools for knowledge discovery. Specifically, the development of modernized data mining capability and systems toxicology should be a future focus, especially methodologies to effectively extract and discover knowledge from very large data sets, to maintain high quality data and provenance information, and to promote use and sharing of public data. Besides the need for developing effective knowledge discovery tools, technology innovations will also impact the application of toxicogenomics in the future. Notably, RNA-Seq is a newly emerging technology for both mapping and quantifying transcriptome. Compared with earlier methods, massively parallel RNA sequencing has vastly increased the throughput of RNA sequencing and allowed global measurement of transcript abundance without the constraint of probes on the chip. The most significant advantage of RNA-Seq is its enabling of the identification of novel biomarkers (alternative splicing, mutation, isoform-specific expression, non-coding RNA, etc.). In addition, RNA-Seq will generate an order of magnitude more data per experiment. Employing the required computational power and data analysis techniques will require corresponding infrastructure designs incorporating massive storage and high internal bandwidth. With improvement in technologies and analysis algorithms, microarray and RNA-Seq combined holds great promise to reveal deeper insights into understanding fundamentals of gene expression variations. While toxicogenomics is not expected to replace traditional toxicological methods, the hope is that it will aid in both the earlier elimination of toxic compounds from the drug pipeline and the discovery of new pathways of toxicity. If so, toxicogenomics and traditional toxicology will perform synergistically in spawning and testing hypotheses, and allow a complementary weight-of-evidence approach.

References

[1] FDA, <http://www.fda.gov/downloads/AboutFDA/Transparency/Basics/UCM247465.pdf>; 2012 [accessed 08.01.13].

[2] Suresh S. Moving toward global science. Science 2011;333(6044):802.

[3] Blomme E, Semizarov D, Blomme E, editors. Genomics in drug discovery and development. Basic principles of toxicology in drug discovery and development. New Jersey: John Wiley & Sons, Inc.; 2009.

[4] Gwathmey JK, Tsaioun K, Hajjar RJ. Cardionomics: a new integrative approach for screening cardiotoxicity of drug candidates. Expert Opin Drug Metab Toxicol 2009;5(6):647–60.

[5] Redfern WS, et al. Safety pharmacology – a progressive approach. Fundam Clin Pharmacol 2002;16(3):161–73.

[6] Westhouse RA. Safety assessment considerations and strategies for targeted small molecule cancer therapeutics in drug discovery. Toxicol Pathol 2010;38(1):165–8.

[7] Kola I, Landis J. Can the pharmaceutical industry reduce attrition rates? Nat Rev Drug Discov 2004; 3(8):711–15.

[8] Khromava A, et al. Manufacturers' postmarketing safety surveillance of influenza vaccine exposure in pregnancy. Am J Obstet Gynecol 2012;207(3):S52–6.

[9] Sakai R, et al. Time-dependent increased risk for serious infection from continuous use of tumor necrosis factor antagonists over three years in patients with rheumatoid arthritis. Arthritis Care Res 2012;64(8):1125–34.

[10] Williams D. Monitoring medicines use: the role of the clinical pharmacologist. Br J Clin Pharmacol 2012;74(4):685–90.

[11] Wills Q, Mitchell C. Toxicogenomics in drug discovery and development – making an impact. Atla Altern Lab Anim 2009;37:33–7.

[12] Nuwaysir EF, Bittner M, Trent J, Barrett JC, Afshari CA. Microarrays and toxicology: the advent of toxicogenomics. Mol Carcinog 1999;24(3):153–9.

[13] NAC, Committee on Applications of Toxicogenomics Technologies to Predictive Toxicology, 'Applications of Toxicogenomic Technologies to Predictive Toxicology and Risk Assessment'. 2007.

[14] Gallagher WM, Tweats D, Koenig J. Omic profiling for drug safety assessment: current trends and public-private partnerships. Drug Discov Today 2009;14(7–8):337–42.

[15] Jacobs A. An FDA perspective on the nonclinical use of the X-Omics technologies and the safety of new drugs. Toxicol Lett 2009;186(1):32–5.

[16] Thomas RS, et al. Application of genomics to toxicology research. Environ Health Perspect 2002;110:919–23.

[17] Frueh FW, Huang SM, Lesko LJ. Regulatory acceptance of toxicogenomics data. Environ Health Perspect 2004;112(12):A663–4.

[18] Kondo C, et al. Predictive genomic biomarkers for drug-induced nephrotoxicity in mice. J Toxicol Sci 2012;37(4):723–37.

[19] Koufaris C, et al. Hepatic MicroRNA profiles offer predictive and mechanistic insights after exposure to genotoxic and epigenetic hepatocarcinogens. Toxicol Sci 2012;128(2):532–43.

[20] Raghavan N, et al. A linear prediction rule based on ensemble classifiers for non-genotoxic carcinogenicity. Statist Biopharm Res 2012;4(2):185–93.

[21] Rusyn I, et al. Predictive modeling of chemical hazard by integrating numerical descriptors of chemical structures and short-term toxicity assay data. Toxicol Sci 2012;127(1):1–9.

[22] Doktorova TY, et al. Comparison of genotoxicant-modified transcriptomic responses in conventional and epigenetically stabilized primary rat hepatocytes with *in vivo* rat liver data. Arch Toxicol 2012;86(11):1703–15.

[23] Jetten MJA, et al. Omics analysis of low dose acetaminophen intake demonstrates novel response pathways in humans. Toxicol Appl Pharmacol 2012;259(3):320–8.

[24] Liao MY, Liu HG. Gene expression profiling of nephrotoxicity from copper nanoparticles in rats after repeated oral administration. Environ Toxicol Pharmacol 2012;34(1):67–80.

[25] Faqi A. A comprehensive guide to toxicology in preclinical drug development. London: Elsevier; 2013.

[26] Hwang MS, et al. Application of toxicogenomic technology for the improvement of risk assessment. Mol Cell Toxicol 2008;4(3):260–6.

[27] Wills Q. SimuGen Ltd: reliable, early prediction of drug toxicity with toxicogenomics, human cell culture and computational models. Pharmacogenomics 2007;8(8):1081–4.

[28] Amir-Aslani A. Toxicogenomic predictive modeling: emerging opportunities for more efficient drug discovery and development. Technol Forecast Soc Change 2008;75(7):905–32.

[29] Suter L, et al. EU Framework 6 Project: predictive toxicology (PredTox) – overview and outcome. Toxicol Appl Pharmacol 2011;252(2):73–84.

[30] Thomas CE, Will Y. The impact of assay technology as applied to safety assessment in reducing compound attrition in drug discovery. Expert Opin Drug Discov 2012;7(2):109–22.

[31] Poma A, Di Giorgio ML. Toxicogenomics to improve comprehension of the mechanisms underlying responses of *in vitro* and *in vivo* systems to nanomaterials: a review. Curr Genomics 2008;9(8):571–85.

[32] Aardema MJ, MacGregor JT. Toxicology and genetic toxicology in the new era of 'toxicogenomics': impact of '-omics' technologies. Mutat Res Fund Mol Mech Mutagen 2002;499(1):13–25.

[33] Chen MJ, Zhang M, Borlak J, Tong W, A decade of toxicogenomic research and its contribution to toxicological science. Toxicol Sci 2012;130(2):217–28.

[34] Beger RD, Colatsky T. Metabolomics data and the biomarker qualification process. Metabolomics 2012;8(1):2–7.

[35] Delles C, Neisius U, Carty DM. Proteomics in hypertension and other cardiovascular diseases. Ann Med 2012;44:S55–64.

[36] Schrattenholz A, et al. Protein biomarkers for *in vitro* testing of toxicology. Mutat Res Genetic Toxicol Environ Mutagen 2012;746(2):113–23.

[37] Quackenbush J. Microarrays – guilt by association. Science 2003;302(5643):240–1.

[38] Lujambio A, Lowe SW. The microcosmos of cancer. Nature 2012;482(7385):347–55.

[39] Lindow M, Kauppinen S. Discovering the first microRNA-targeted drug. J Cell Biol 2012;199(3):407–12.

[40] Mendrick DL. Transcriptional profiling to identify biomarkers of disease and drug response. Pharmacogenomics 2011;12(2):235–49.

[41] Chen BS, Li CW. Analyzing microarray data in drug discovery using systems biology. Expert Opin Drug Discov 2007;2(5):755–68.

[42] Ma HC, Horiuchi KY. Chemical microarray: a new tool for drug screening and discovery. Drug Discov Today 2006;11(13–14):661–8.

[43] Petersen D, et al. Three microarray platforms: an analysis of their concordance in profiling gene expression. BMC Genomics 2005;6(1):63.

[44] Hardiman G. Microarray platforms – comparisons and contrasts. Future Med 2004;5(5):487–502.

[45] Ju W, et al. Identification of genes with differential expression in chemoresistant epithelial ovarian cancer using high-density oligonucleotide microarrays. Oncol Res 2009;18(2–3):47–56.

[46] Zidek N, et al. Acute hepatotoxicity: a predictive model based on focused Illumina microarrays. Toxicol Sci 2007;99(1):289–302.

[47] Schulze A, Downward J. Navigating gene expression using microarrays – a technology review. Nat Cell Biol 2001;3(8):E190–5.

[48] Edwards HD, Nagappayya SK, Pohl NLB. Probing the limitations of the fluorous content for tag-mediated microarray formation. Chem Commun 2012;48(4):510–12.

[49] Khouja MH, et al. Limitations of tissue microarrays compared with whole tissue sections in survival analysis. Oncol Lett 2010;1(5):827–31.

[50] Tanase CP, Albulescu R, Neagu M. Application of 3D hydrogel microarrays in molecular diagnostics: advantages and limitations. Expert Rev Mol Diagn 2011;11(5):461–4.

[51] Weisenberg JLZ, et al. Diagnostic yield and limitations of chromosomal microarray: a retrospective chart review. Ann Neurol 2008;64:S101.

[52] Lin LL, Huang HC, Juan HF. Revealing the molecular mechanism of gastric cancer marker Annexin A4 in cancer cell proliferation using exon arrays. Plos One 2012;7(9):e44615.

[53] Raghavachari N, et al. A systematic comparison and evaluation of high density exon arrays and RNA-Seq technology used to unravel the peripheral blood transcriptome of sickle cell disease. BMC Med Genomics 2012;5:28.

[54] Mockler TC, et al. Applications of DNA tiling arrays for whole-genome analysis. Genomics 2005;85(5):655.

[55] Fiesel FC, et al. TDP-43 regulates global translational yield by splicing of exon junction complex component SKAR. Nucleic Acids Res 2012;40(6):2668–82.

[56] Liu S, et al. A comparison of RNA-Seq and high-density exon array for detecting differential gene expression between closely related species. Nucleic Acids Res 2011;39(2):578–88.

[57] Sakarya O, et al. RNA-Seq mapping and detection of gene fusions with a suffix array algorithm. PLoS Comput Biol 2012;8(4):e1002464.

[58] Seok J, et al. JETTA: junction and exon toolkits for transcriptome analysis. Bioinformatics 2012;28(9):1274–5.

[59] Bi K, Vanderpool D, Singhal S, Linderoth T, Moritz C, Good JM. Transcriptome-based exon capture enables highly cost-effective comparative genomic data collection at moderate evolutionary scales. BMC Genomics 2012;13:403 (17 August 2012).

[60] Kapetis D, et al. AMDA 2.13: a major update for automated cross-platform microarray data analysis. Biotechniques 2012;53(1):33.

[61] Gellert P, Ponomareva Y, Braun T, Uchida S. Noncoder: a web interface for exon array-based detection of long non-coding RNAs. Nucleic Acids Res 2013;41(1):e20.

[62] David L, et al. A high-resolution map of transcription in the yeast genome. Proc Natl Acad Sci USA 2006;103(14):5320–5.

[63] Cheng J, et al. Transcriptional maps of 10 human chromosomes at 5-nucleotide resolution. Science 2005;308(5725):1149–54.

[64] Clark TA, Sugnet CW, Ares M. Genomewide analysis of mRNA processing in yeast using splicing-specific microarrays. Science 2002;296(5569):907–10.

[65] Morra L, et al. Characterization of periostin isoform pattern in non-small cell lung cancer. Lung Cancer 2012;76(2):183–90.

[66] Yeo GWM. Splicing regulators: targets and drugs. Genome Biol 2005;6:12.

[67] Farber CR. Systems-level analysis of genome-wide association data. G3 Genes Genomes Genet 2013;3(1):119–29.

[68] Leveque N, Renois F, Andreoletti L. The microarray technology: facts and controversies. Clin Microbiol Infect 2013;19(1):10–14.

[69] Licatalosi DD, Darnell RB. Applications of next-generation sequencing RNA processing and its regulation: global insights into biological networks. Nat Rev Genet 2010;11(1):75–87.

[70] Lister R, Gregory BD, Ecker JR. Next is now: new technologies for sequencing of genomes, transcriptomes, and beyond. Curr Opin Plant Biol 2009;12(2):107–18.

[71] Morozova O, Hirst M, Marra MA. Applications of new sequencing technologies for transcriptome analysis. Annu Rev Genomics Hum Genet 2009:135–51.

[72] Smith AM, et al. Quantitative phenotyping via deep barcode sequencing. Genome Res 2009;19(10):1836–42.

[73] Krupp M, et al. RNA-Seq Atlas – a reference database for gene expression profiling in normal tissue by next-generation sequencing. Bioinformatics 2012;28(8):1184–5.

[74] Li JJ, et al. Sparse linear modeling of next-generation mRNA sequencing (RNA-Seq) data for isoform discovery and abundance estimation. Proc Natl Acad Sci USA 2011;108(50):19867–72.

[75] Bentley DR, et al. Accurate whole human genome sequencing using reversible terminator chemistry. Nature 2008;456(7218):53–9.

[76] Ardlie K, et al. Analysis of RNAStable treated samples by the illumina DASL assay and illumina RNA sequencing. J Mol Diagn 2012;14(6):637.

[77] Chelliserry M, et al. Can next generation sequencing replace sanger sequencing? – A review of the Illumina cystic fibrosis diagnostic test on the MiSeqDx instrument. J Mol Diagn 2012;14(6):651.

[78] Williams LJS, et al. Paired-end sequencing of Fosmid libraries by Illumina. Genome Res 2012;22(11):2241–9.

[79] Marioni JC, et al. RNA-Seq: an assessment of technical reproducibility and comparison with gene expression arrays. Genome Res 2008;18(9):1509–17.

[80] Wang Z, Gerstein M, Snyder M. RNA-Seq: a revolutionary tool for transcriptomics. Nat Rev Genet 2009;10(1):57–63.

[81] Iancu OD, et al. Utilizing RNA-Seq data for *de novo* coexpression network inference. Bioinformatics 2012;28(12):1592–7.

[82] de Magalhaes JP, Finch CE, Janssens G. Next-generation sequencing in aging research: emerging applications, problems, pitfalls and possible solutions. Ageing Res Rev 2010;9(3):315–23.

[83] Mutz KO, et al. Transcriptome analysis using next-generation sequencing. Curr Opin Biotechnol 2013;24(1):22–30.

[84] Werner T. Next generation sequencing in functional genomics. Brief Bioinform 2010;11(5):499–511.

[85] Olsen LCB, Faergeman NJ. Chemical genomics and emerging DNA technologies in the identification of drug mechanisms and drug targets. Curr Top Med Chem 2012;12(12):1331–45.

[86] Ozsolak F. Third-generation sequencing techniques and applications to drug discovery. Expert Opin Drug Discov 2012;7(3):231–43.

[87] Woollard PM, et al. The application of next-generation sequencing technologies to drug discovery and development. Drug Discov Today 2011;16(11–12):512–19.

[88] Jenne A, et al. Drug profiling and biomarker discovery using mass spectrometry-based proteomics technologies. J Biotechnol 2010;150:S29.

[89] Prunotto M, et al. Urinary proteomics and drug discovery in chronic kidney disease: a new perspective. J Proteome Res 2011;10(1):126–32.

[90] Komatsu S, et al. Comprehensive analysis of endoplasmic reticulum-enriched fraction in root tips of soybean under flooding stress using proteomics techniques. J Proteomics 2012;77:531–60.

[91] Steinberger B, et al. Comparison of gel-based phosphoproteomic approaches to analyse scarce oviductal epithelial cell samples. Proteomics 2013;13(1):12–16.

[92] Van Summeren A, et al. Proteomics in the search for mechanisms and biomarkers of drug-induced hepatotoxicity. Toxicol *In Vitro* 2012;26(3):373–85.

[93] Collins BC, et al. Development of a pharmaceutical hepatotoxicity biomarker panel using a discovery to targeted proteomics approach. Mol Cell Proteomics 2012;11(8):394–410.

[94] Singh S, et al. Omics in mechanistic and predictive toxicology. Toxicol Mech Methods 2010;20(7):355–62.

[95] Miao Q, Zhang CC, Kast J. Chemical proteomics and its impact on the drug discovery process. Expert Rev Proteomics 2012;9(3):281–91.

[96] Beck M, et al. The quantitative proteome of a human cell line. Mol Syst Biol 2011;7:549.

[97] Beyoglu D, Idle JR. Metabolomics and its potential in drug development. Biochem Pharmacol 2013;85(1):12–20.

[98] Robertson DG, Reily MD. The current status of metabolomics in drug discovery and development. Drug Dev Res 2012;73(8):535–46.

[99] D'Alessandro A, Zolla L. Metabolomics and cancer drug discovery: let the cells do the talking. Drug Discov Today 2012;17(1–2):3–9.

[100] Wilcoxen KM, et al. Practical metabolomics in drug discovery. Expert Opin Drug Discov 2010;5(3):249–63.

[101] Zhang T, et al. Recent progress on bioinformatics, functional genomics, and metabolomics research of cytochrome P450 and its impact on drug discovery. Curr Top Med Chem 2012;12(12):1346–55.

[102] Eckhart AD, Beebe K, Milburn M. Metabolomics as a key integrator for 'omic' advancement of personalized medicine and future therapies. Clin Transl Sci 2012;5(3):285–8.

[103] Beger RD, Sun JC, Schnackenberg LK. Metabolomic approaches for discovering biomarkers of drug-induced hepatotoxicity and nephrotoxicity. Toxicol Appl Pharmacol 2010;243(2):154–66.

[104] Kumar BS, et al. Discovery of common urinary biomarkers for hepatotoxicity induced by carbon tetrachloride, acetaminophen and methotrexate by mass spectrometry-based metabolomics. J Appl Toxicol 2012;32(7):505–20.

[105] Lee BM, et al. Toxicogeno-metabolomics approach for the discovery of nephrotoxicity biomarkers. Toxicol Lett 2011;205:S215.

[106] Clayton TA, et al. Pharmaco-metabonomic phenotyping and personalized drug treatment. Nature 2006;440(7087):1073–7.

GENOMIC BIOMARKERS FOR PHARMACEUTICAL DEVELOPMENT

[107] Winnike JH, et al. Use of pharmaco-metabolomics for early prediction of acetaminophen-induced hepatotoxicity in humans. Clin Pharmacol Ther 2010;88(1):45–51.

[108] FDA guidance for industry, E16 biomarkers related to drug or biotechnology product development' context, structure, and format of qualification submissions. August 2011 ICH.

[109] Scannell JW, et al. Diagnosing the decline in pharmaceutical R&D efficiency. Nat Rev Drug Discov 2012;11(3):191–200.

[110] Chen MJ, et al. A decade of toxicogenomic research and its contribution to toxicological science. Toxicol Sci 2012;130(2):217–28.

[111] Choudhuri S. Looking back to the future: from the development of the gene concept to toxicogenomics. Toxicol Mech Methods 2009;19(4):263–77.

[112] Wei T, Li SY. Development of genomics-based gene expression signature biomarkers in oncology and toxicology to facilitate drug discovery and translational medicine. Curr Bioinformatics 2010;5(2):109–17.

[113] Ryan TP, Stevens JL, Thomas CE. Strategic applications of toxicogenomics in early drug discovery. Curr Opin Pharmacol 2008;8(5):654–60.

[114] Stevens J. Strategic application of toxicogenomics in early drug discovery. Drug Metab Rev 2010;42:3–4.

[115] Bates S. The role of gene expression profiling in drug discovery. Curr Opin Pharmacol 2011;11(5):549–56.

[116] Sirota M, et al. Discovery and preclinical validation of drug indications using compendia of public gene expression data (vol 3, 96ra77, 2011). Sci Transl Med 2011;3(102):96ra77.

[117] Lord PG, Nie A, McMillian M. The evolution of gene expression studies in drug safety assessment (vol 16, pg 51, 2006). Toxicol Mech Methods 2006;16(4):241.

[118] Ulrich R. Applications of gene expression profiling to evaluating drug metabolism and safety. Drug Metab Rev 2005;37:7.

[119] Harrill AH, Ross PK, Gatti DM, Threadgill DW, Rusyn I. Population-based discovery of toxicogenomics biomarkers for hepatotoxicity using a laboratory strain diversity panel. Toxicol Sci 2009;110(1):235–43.

[120] Kiyosawa N, Ando Y, Manabe S, Yamoto T. Toxicogenomic biomarkers for liver toxicity. J Toxicol Pathol 2009;22(1):35–52.

[121] Uehara T, et al. The japanese toxicogenomics project: application of toxicogenomics. Mol Nutr Food Res 2010;54(2):218–27.

[122] Zhang M, Chen MJ, Tong WD. Is toxicogenomics a more reliable and sensitive biomarker than conventional indicators from rats to predict drug-induced liver injury in humans? Chem Res Toxicol 2012;25:1.

[123] Huang J, Shi W, et al. Genomic indicators in the blood predict drug-induced liver injury. Pharmacogenomics J 2010;10:353–63.

[124] Kerns RTB, Bushel PR. The impact of classification of interest on predictive toxicogenomics. Front Genet 2012;3(14).

[125] Chen MJ, et al. FDA-approved drug labeling for the study of drug-induced liver injury. Drug Discov Today 2011;16(15–16):697–703.

[126] Stine JG, Lewis JH. Drug-induced liver injury: a summary of recent advances. Expert Opin Drug Metab Toxicol 2011;7(7):875–90.

[127] Lucena MI, et al. Phenotypic characterization of idiosyncratic drug-induced liver injury: the influence of age and sex. Hepatology 2009;49(6):2001–9.

[128] Hou FQ, Zeng Z, Wang GQ. Hospital admissions for drug-induced liver injury: clinical features, therapy, and outcomes. Cell Biochem Biophys 2012;64(2):77–83.

[129] Carey EJ, et al. Inpatient admissions for drug-induced liver injury: results from a single center. Dig Dis Sci 2008;53(7):1977–82.

[130] Lee WM. Medical progress: drug-induced hepatotoxicity. N Engl J Med 2003;349(5):474–85.

[131] Reuben A, Koch DG, Lee WM. Drug-induced acute liver failure: results of a US multicenter, prospective study. Hepatology 2010;52(6):2065–76.

[132] McDonnell ME, Braverman LE. Drug-related hepatotoxicity. N Engl J Med 2006;354(20):2191.

[133] Navarro VJ, Senior JR. Current concepts – Drug-related hepatotoxicity. N Engl J Med 2006;354(7):731–9.

[134] Lee WM. Recent developments in acute liver failure. Best Pract Res Clin Gastroenterol 2012;26(1):3–16.

[135] Lee WM. Acute liver failure. Semin Respir Crit Care Med 2012;33(1):36–45.

[136] FDA guidance for industry, Drug-induced liver injury: premarketing clinical evaluation. July 2009 Drug Safety.

[137] Abboud G, Kaplowitz N. Drug-induced liver injury. Drug Saf 2007;30(4):277–94.

[138] Hussaini SH, Farrington EA. Idiosyncratic drug-induced liver injury: an overview. Expert Opin Drug Saf 2007;6(6):673–84.

[139] Eun JW, et al. Discriminating the molecular basis of hepatotoxicity using the large-scale characteristic molecular signatures of toxicants by expression profiling analysis. Toxicology 2008;249(2–3):176–83.

[140] Low Y, et al. Predicting drug-induced hepatotoxicity using QSAR and toxicogenomics approaches. Chem Res Toxicol 2011;24(8):1251–62.

[141] Parman T, et al. Toxicogenomics and metabolomics of pentamethylchromanol (PMCol)-induced hepatotoxicity. Toxicol Sci 2011;124(2):487–501.

[142] Zhao Q, et al. Epigenetic modifications in hepatic stellate cells contribute to liver fibrosis. Tohoku J Exp Med 2013;229(1):35–43.

[143] Xie GH, et al. Hedgehog signaling regulates liver sinusoidal endothelial cell capillarisation. Gut 2013;62(2):299–309.

[144] Loi P, et al. Interferon regulatory factor 3 deficiency leads to interleukin-17-mediated liver ischemia-reperfusion injury. Hepatology 2013;57(1):351–61.

[145] Hawkins MT, Lewis JH. Latest advances in predicting DILI in human subjects: focus on biomarkers. Expert Opin Drug Metab Toxicol 2012;8(12):1521–30.

[146] Zheng JL, et al. Assessment of subclinical, toxicant-induced hepatic gene expression profiles after low-dose, short-term exposures in mice. Regul Toxicol Pharmacol 2011;60(1):54–72.

[147] Kiyosawa N, et al. Toxicogenomic biomarkers for liver toxicity. J Toxicol Pathol 2009;22(1):35–52.

[148] Yang X, Salminen W, et al. Current and emerging biomarkers of hepatotoxicity. Curr Biomarker Find 2012;2:43–55.

[149] Kuehn BM. FDA focuses on drugs and liver damage labeling and other changes for acetaminophen. JAMA 2009;302(4):369–71.

[150] Bushel PR, et al. Blood gene expression signatures predict exposure levels. Proc Natl Acad Sci USA 2007;104(46):18211–16.

[151] Johansson I, Ingelman-Sundberg M. Genetic polymorphism and toxicology – with emphasis on cytochrome P450. Toxicol Sci 2011;120(1):1–13.

[152] Przybylak KR, Cronin MTD. *In silico* models for drug-induced liver injury – current status. Expert Opin Drug Metab Toxicol 2012;8(2):201–17.

[153] ICH Topic S4, duration of chronic toxicity testing in animals (rodent and non rodent toxicity testing), May 1999 CPMP/ICH?300/95.

[154] Arcangeli S, et al. Updated results and patterns of failure in a randomized hypofractionation trial for high-risk prostate cancer. Int J Radiat Oncol Biol Phys 2012;84(5):1172–8.

[155] LeCluyse EL, et al. Organotypic liver culture models: meeting current challenges in toxicity testing. Crit Rev Toxicol 2012;42(6):501–48.

[156] Landis GN, et al. Similar gene expression patterns characterize aging and oxidative stress in *Drosophila melanogaster*. Proc Natl Acad Sci USA 2004;101(20):7663–8.

[157] Bol D, Ebner R. Gene expression profiling in the discovery, optimization and development of novel drugs: one universal screening platform. Pharmacogenomics 2006;7(2):227–35.

[158] Hoflack JC, Roth AB, Suter L. Toxicogenomics: a predictive tool in toxicology and drug development. Drug Metab Rev 2010;42:4.

[159] Nam SW. Expression toxicogenomics in predictive toxicology and risk assessment. Mol Cell Toxicol 2009; 5(3):45.

[160] Williams-Devane CR, Wolf MA, Richard AM. Toward a public toxicogenomics capability for supporting predictive toxicology: survey of current resources and chemical indexing of experiments in GEO and ArrayExpress. Toxicol Sci 2009;109(2):358–71.

[161] Vanhaecke T, et al. EU research activities in alternative testing strategies: current status and future perspectives. Arch Toxicol 2009;83(12):1037–42.

[162] Yamamoto K, et al. Effect of globin digest on the liver injury and hepatic gene expression profile in galactosamine-induced liver injury in SD rats. Life Sci 2011;88(15–16):701–12.

[163] Yang Q, et al. Dynamics of short-term gene expression profiling in liver following thermal injury. J Surg Res 2012;176(2):549–58.

[164] Suh SK, et al. Gene expression profiling of acetaminophen induced hepatotoxicity in mice. Mol Cell Toxicol 2006;2(4):236–43.

[165] Upadhyay G, et al. Involvement of multiple molecular events in pyrogallol-induced hepatotoxicity and sily-marin-mediated protection: evidence from gene expression profiles. Food Chem Toxicol 2010;48(6):1660–70.

[166] Yun JW, et al. Predose blood gene expression profiles might identify the individuals susceptible to carbon tetrachloride-induced hepatotoxicity. Toxicol Sci 2010;115(1):12–21.

[167] McMillian M, Nie A, Parker JB, Leone A, Kemmerer M, Bryant S, et al. Drug-induced oxidative stress in rat liver from a toxicogenomics perspective. Toxicol Appl Pharmacol 2005;207:S171–8.

[168] Seifert O, Matussek A, Sjögren F, Geffers R, Anderson CD. Gene expression profiling of macrophages: impli-cations for an immunosuppressive effect of dissolucytotic gold ions. J Inflamm (London) 2012;9:43.

[169] Jiang Y, et al. Diagnosis of drug-induced renal tubular toxicity using global gene expression profiles. J Transl Med 2007;5:47.

[170] Kim YS. Changes of gene expression profiles in stable renal tubule epithelial cell lines as biomarkers of drug-specific toxicities. Toxicol Sci 2003;72:244.

[171] NAC, Application of toxicogenomics technologies to predictive toxicology and risk assessment. The National Academic Press. 2007.

[172] Ellinger-Ziegelbauer H, et al. Prediction of a carcinogenic potential of rat hepatocarcinogens using toxicog-enomics analysis of short-term *in vivo* studies. Mutat Res Fund Mol Mech Mutagen 2008;637(1–2):23–39.

[173] Daston GP. Gene expression, dose-response, and phenotypic anchoring: applications for toxicogenomics in risk assessment. Toxicol Sci 2008;105(2):233–4.

[174] Waring WS, Moonie A. Earlier recognition of nephrotoxicity using novel biomarkers of acute kidney injury. Clin Toxicol 2011;49(8):720–8.

[175] Khor TO, Ibrahim S, Kong ANT. Toxicogenomics in drug discovery and drug development: potential appli-cations and future challenges. Pharm Res 2006;23(8):1659–64.

[176] Park HJ, et al. Identification of biomarkers of chemically induced hepatocarcinogenesis in rasH2 mice by tox-icogenomic analysis. Arch Toxicol 2011;85(12):1627–40.

[177] Wang T, et al. Investigation of correlation among safety biomarkers in serum, histopathological examination, and toxicogenomics. Int J Toxicol 2011;30(3):300–12.

[178] Harmey D, Griffin PR, Kenny PJ. Development of novel pharmacotherapeutics for tobacco dependence: pro-gress and future directions. Nicotine Tob Res 2012;14(11):1300–18.

[179] Simo R, Hernandez C. Neurodegeneration is an early event in diabetic retinopathy: therapeutic implications. Br J Ophthalmol 2012;96(10):1285–90.

[180] Zhou LM, et al. Pharmacokinetic and pharmacodynamic interaction of Danshen-Gegen extract with warfarin and aspirin. J Ethnopharmacol 2012;143(2):648–55.

[181] Ban JY, et al. Gene expression profiles regulated by Hspa1b in MPTP-induced dopaminergic neurotoxicity using knockout mice. Mol Cell Toxicol 2012;8(3):281–7.

[182] Bu Q, et al. Transcriptome analysis of long non-coding RNAs of the nucleus accumbens in cocaine-condi-tioned mice. J Neurochem 2012;123(5):790–9.

[183] Meli L, et al. Influence of a three-dimensional, microarray environment on human Cell culture in drug screening systems. Biomaterials 2012;33(35):9087–96.

[184] Fery Y, Mueller SO, Schrenk D. Co-regulation of CYP3A and PXR gene expression in rat and human hepatoma cells. Naunyn Schmiedebergs Arch Pharmacol 2007;375:73.

[185] Zhang WX, et al. Co-regulation of mRNA level of UDP glucuronosyltransferase 1A9 and Multi-drug resist-ance protein 2 in Chinese human liver. Clin Chim Acta 2010;411(1–2):119–21.

[186] Au JS, Navarro VJ, Rossi S. Review article: drug-induced liver injury – its pathophysiology and evolving diagnostic tools. Aliment Pharmacol Ther 2011;34(1):11–20.

[187] Mori Y, et al. Identification of potential genomic biomarkers for early detection of chemically induced cardio-toxicity in rats. Toxicology 2010;271(1–2):36–44.

[188] Pallet N, Legendre C. Deciphering calcineurin inhibitor nephrotoxicity: a pharmacological approach. Pharmacogenomics 2010;11(10):1491–501.

[189] Thomas R, et al. Using bioinformatic approaches to identify pathways targeted by human leukemogens. Int J Environ Res Public Health 2012;9(7):2479–503.

[190] Waring JF, et al. Microarray analysis of hepatotoxins *in vitro* reveals a correlation between gene expression profiles and mechanisms of toxicity. Toxicol Lett 2001;120(1–3):359–68.

[191] Cui YX, Paules RS. Use of transcriptomics in understanding mechanisms of drug-induced toxicity. Pharmacogenomics 2010;11(4):573–85.

[192] Sawada H, Takami K, Asahi S. A toxicogenomic approach to drug-induced phospholipidosis: analysis of its induction mechanism and establishment of a novel *in vitro* screening system. Toxicol Sci 2005;83(2):282–92.

[193] Sahu SC, editor. Toxicogenomics: a powerful tool for toxicity assessment. UK: Wiley; 2008.

[194] Cunningham ML, Lehman-McKeeman L. Applying toxicogenomics in mechanistic and predictive toxicology. Toxicol Sci 2005;83(2):205–6.

[195] Burczynski ME, et al. Toxicogenomics-based discrimination of toxic mechanism in HepG2 human hepatoma cells. Toxicol Sci 2000;58(2):399–415.

[196] Bulera SJ, et al. RNA expression in the early characterization of hepatotoxicants in wister rats by high-density DNA microarrays. Hepatology 2001;33(5):1239–58.

[197] Minotti G, et al. Anthracyclines: molecular advances and pharmacologic developments in antitumor activity and cardiotoxicity. Pharmacol Rev 2004;56(2):185–229.

[198] IOM, Emerging Safety Science: Workshop Summary. Chapter 4: Screening Technologies II: Toxicogenomics <http://www.ncbi.nlm.nih.gov/books/NBK4060/>; 2008 [accessed 21.01.13].

[199] Raman T, et al. Quality control in microarray assessment of gene expression in human airway epithelium. BMC Genomics 2009;10:493.

[200] Thompson KL, et al. Characterization of the effect of sample quality on high density oligonucleotide microarray data using progressively degraded rat liver RNA. BMC Biotechnol 2007;7:57.

[201] Archer KJ, et al. Application of a correlation correction factor in a microarray cross-platform reproducibility study. BMC Bioinformatics 2007;8:447.

[202] Chen JJ, et al. Reproducibility of microarray data: a further analysis of microarray quality control (MAQC) data. BMC Bioinformatics 2007;8:1–14.

[203] Shi LM, et al. The balance of reproducibility, sensitivity, and specificity of lists of differentially expressed genes in microarray studies. BMC Bioinformatics 2008;9(Suppl. 9):S10.

[204] Kadota K, Shimizu K. Evaluating methods for ranking differentially expressed genes applied to microArray quality control data. BMC Bioinformatics 2011;12:227.

[205] Kurose K, et al. Quality requirements for genomic dna preparations and storage conditions for a high-density oligonucleotide microarray. Biol Pharm Bull 2012;35(10):1846–8.

[206] McCall MN, et al. Assessing affymetrix GeneChip microarray quality. BMC Bioinformatics 2011;12:137.

[207] Schmidberger M, Vicedo E, Mansmann U. Empirical study for the agreement between statistical methods in quality assessment and control of microarray data. Comput Stat 2011;26(2):259–77.

[208] Casciano DA, Woodcock J. Empowering microarrays in the regulatory setting. Nat Biotechnol 2006;24(9):1103.

[209] <http://www.fda.gov/ScienceResearch/BioinformaticsTools/MicroarrayQualityControlProject/>. MAQC-II. FDA [accessed 12.01.13].

[210] Shi LM, et al. The microarray quality control (MAQC)-IIII study of common practices for the development and validation of microarray-based predictive models. Nat Biotechnol 2010;28(8):827-U109.

[211] Shi LM, et al. The microarray quality control (MAQC) project shows inter- and intraplatform reproducibility of gene expression measurements. Nat Biotechnol 2006;24(9):1151–61.

[212] Tong WD, et al. Evaluation of external RNA controls for the assessment of microarray performance. Nat Biotechnol 2006;24(9):1132–9.

[213] Fu X, et al. Data governance in predictive toxicology: a review. J Cheminformatics 2011;3(1):24.

[214] Judson R. Public databases supporting computational toxicology. J Toxicol Environ Health B Crit Rev 2010;13(2–4):218–31.

[215] Judson RS, et al. Aggregating data for computational toxicology applications: the US environmental protection agency (EPA) aggregated computational toxicology resource (ACToR) system. Int J Mol Sci 2012;13(2):1805–31.

[216] Davis AP, et al. The comparative toxicogenomics database: update 2011. Nucleic Acids Res 2011;39:D1067–D1072.

[217] Davis AP, et al. MEDIC: a practical disease vocabulary used at the comparative toxicogenomics database. Database 2012.

[218] Uehara T, et al. Prediction model of potential hepatocarcinogenicity of rat hepatocarcinogens using a large-scale toxicogenomics database. Toxicol Appl Pharmacol 2011;255(3):297–306.

[219] Ganter B, et al. Development of a large-scale chemogenomics database to improve drug candidate selection and to understand mechanisms of chemical toxicity and action. J Biotechnol 2005;119(3):219–44.

[220] Entelos. <www.entelos.com>.

[221] Yokochi S, et al. An anti-inflammatory drug, propagermanium, may target GPI-anchored proteins associated with an MCP-1 receptor, CCR2. J Interferon Cytokine Res 2001;21(6):389–98.

[222] Kim NY, et al. Functional role of phospholipase d (PLD) in di(2-ethylhexyl) phthalate-induced hepatotoxicity in Sprague-Dawley rats. J Toxicol Environ Health A Curr Issues 2010;73(21–22):1560–9.

[223] Ramdhan D H, Kamijima M. Differential response to trichloroethylene-induced hepatosteatosis in wild-type and PPARα-humanized mice. Environ Health Perspect 2010;118(11):1557–63.

[224] Noriyuki N, et al. Evaluation of DNA microarray results in the toxicogenomics project (TGP) consortium in Japan. J Toxicol Sci 2012;37(4):791–801.

[225] Ellinger-Ziegelbauer H, et al. The enhanced value of combining conventional and 'omics' analyses in early assessment of drug-induced hepatobiliary injury. Toxicol Appl Pharmacol 2011;252(2):97–111.

[226] Matheis KA, et al. Cross-study and cross-omics comparisons of three nephrotoxic compounds reveal mechanistic insights and new candidate biomarkers. Toxicol Appl Pharmacol 2011;252(2):112–22.

[227] Adiamah DA, Handl J, Schwartz JM. Streamlining the construction of large-scale dynamic models using generic kinetic equations. Bioinformatics 2010;26(10):1324–31.

[228] Tra YV, Evans IM. Enhancing interdisciplinary mathematics and biology education: a microarray data analysis course bridging these disciplines. CBE Life Sci Educ 2010;9(3):217–26.

[229] Puri R, et al. The emerging role of plasma lipidomics in cardiovascular drug discovery. Expert Opin Drug Discov 2012;7(1):63–72.

[230] Spyridopoulou KP, et al. Methylene tetrahydrofolate reductase gene polymorphisms and their association with methotrexate toxicity: a meta-analysis. Pharmacogenet Genomics 2012;22(2):117–33.

[231] Wu MH, et al. Identification of drug targets by chemogenomic and metabolomic profiling in yeast. Pharmacogenet Genomics 2012;22(12):877–86.

[232] Castillo S, et al. Algorithms and tools for the preprocessing of LC-MS metabolomics data. Chemometrics Intell Laboratory Syst 2011;108(1):23–32.

[233] Xu G, et al. Improve accuracy and sensibility in glycan structure prediction by matching glycan isotope abundance. Anal Chim Acta 2012;743:80–9.

[234] Avramenko Y, et al. Mining of graphics for information and knowledge retrieval. Comput Chem Eng 2009;33(3):618–27.

[235] Nebel ME, Scheid A, Weinberg F. Random generation of RNA secondary structures according to native distributions. Algorithms Mol Biol 2011;6(1):25.

[236] Coenen F. Data mining: past, present and future. Knowledge Eng Rev 2011;26(1):25–9.

[237] Kumar V, et al., High performance data mining. In: Palma JML, et al., editors. High Performance Computing for Computational Science – Vecpar 2002;2003, pp. 111–25.

[238] Hammann F, Drewe J. Decision tree models for data mining in hit discovery. Expert Opin Drug Discov 2012;7(4):341–52.

[239] Gao J, et al. A graph-based consensus maximization approach for combining multiple supervised and unsupervised models. IEEE Trans Knowledge Data Eng 2013;25(1):15–28.

[240] Lee CH. Unsupervised and supervised learning to evaluate event relatedness based on content mining from social-media streams. Expert Syst Appl 2012;39(18):13338–56.

[241] Osoba O, Kosko B. Noise-enhanced clustering and competitive learning algorithms. Neural Netw 2013;37:132–40.

[242] Ahn B, et al. Increasing splicing site prediction by training gene set based on species. KSII Trans Internet Inform Syst 2012;6(11):2784–99.

[243] Persello C, Bruzzone L. Active learning for domain adaptation in the supervised classification of remote sensing images. IEEE Trans Geosci Remote Sens 2012;50(11):4468–83.

[244] Pinto C, Bridges M, Perez I. A supervised learning approach to the genetic classification of populations. Ann Hum Genet 2012;76:413.

[245] Wang HR, et al. Supervised class-specific dictionary learning for sparse modeling in action recognition. Pattern Recognit 2012;45(11):3902–11.

[246] Botelho F, Davis A. Stability behavior for unsupervised learning. Physica D 2013;243(1):111–15.

[247] Guzzi PH, et al. DMET-Analyzer: automatic analysis of affymetrix DMET Data. BMC Bioinformatics 2012;13:258.

[248] Kayala MA, Baldi P. Cyber-T web server: differential analysis of high-throughput data. Nucleic Acids Res 2012;40(W1):W553–9.

[249] Ranagathan S, et al. Advances in translational bioinformatics and population genomics in the Asia-Pacific. BMC Genomics 2012;(Suppl. 7):S1.

[250] Smit-McBride Z, et al. Genomes/genomics/bioinformatics. *In vitro*. Cell Dev Biol Anim 2012;48(7):459.

[251] Sung WK. Bioinformatics applications in genomics. Computer 2012;45(6):57–63.

[252] Carazzolle MF, et al. D-MaPs – DNA-microarray projects: web-based software for multi-platform microarray analysis. Genet Mol Biol 2009;32(3):634–9.

[253] Chen X, et al. Supervised principal component analysis for gene set enrichment of microarray data with continuous or survival outcomes. Bioinformatics 2008;24(21):2474–81.

[254] Lajugie J, Bouhassira EE. GenPlay, a multipurpose genome analyzer and browser. Bioinformatics 2011;27(14):1889–93.

[255] Scharl T, et al. Interactive visualization of clusters in microarray data: an efficient tool for improved metabolic analysis of E. coli. Microb Cell Fact 2009;8:37.

[256] Butte A. The use and analysis of microarray data. Nat Rev Drug Discov 2002;1(12):951–60.

[257] Cano C, et al. Intelligent system for the analysis of microarray data using principal components and estimation of distribution algorithms. Expert Syst Appl 2009;36(3):4654–63.

[258] Pounds S, et al. Integrated analysis of pharmacologic, clinical and SNP microarray data using Projection Onto the Most Interesting Statistical Evidence with Adaptive Permutation Testing. Int J Data Min Bioinform 2011;5(2):143–57.

[259] FDA guidance for industry: E15 definitions for genomics biomarkers, pharmacogenomics, pharmacogenetics, genomic data and sample coding Categories. April 2008 ICH.

Translating Biomarker Discovery into Companion Diagnostics through Validation and Regulatory Consideration

Philip Brohawn, Brandon W. Higgs, Koustubh Ranade, Bahija Jallal, Yihong Yao

MedImmune, LLC, Gaithersburg, Maryland

7.1 INTRODUCTION

Prior to describing strategies for the effective analytical validation and clinical implementation of a companion diagnostic device, it is important to have a clear understanding of the most current regulatory guidance on the subject. Currently there are a number of different companion diagnostic approval regulatory pathways that depend upon the geographical location of implementation and/or development. For example, self certification and obtaining a CE mark is the path in Europe, while in the United States, the Food and Drug Administration (FDA) has a more defined regulatory pathway and approval process for devices. It is widely accepted that the most rigorous of these regulatory pathways is that laid out by the FDA, and that meeting the requirements of FDA premarket approval will likely be sufficient to meet the requirements of other regulatory bodies. For this reason we will examine in greater depth the regulatory path to approval as defined by the FDA.

On July 25, 2011, the FDA's Center for Devices and Radiological Health (CDRH) division released a version of draft guidance on companion diagnostics to solicit feedback from industry [1]. This draft guidance followed July 2007 draft guidance on *In vitro* Diagnostic Multivariate Index Assays [2], and an even earlier April 2005 draft Drug Diagnostic Co-Development Concept Paper [3]. As the 'Draft Guidance for Industry and Food and Drug Administration Staff – *In vitro* Companion Diagnostic Devices' is the most updated guidance,

Y. Yao, B. Jallal, K. Ranade (Eds): Genomic Biomarkers for Pharmaceutical Development.
DOI: http://dx.doi.org/10.1016/B978-0-12-397336-8.00007-0

157

this chapter will review the contents of this document. The full guidance document can be reviewed at the following link: http://www.fda.gov/medicaldevices/deviceregulationand-guidance/guidancedocuments/ucm262292.htm (accessed December 2012). The guidance is intended to cover all sponsors for whom a device is required to ensure the safe and efficacious use of a therapeutic, as well as those choosing to develop the device in association with an intended therapy. With the evolution of personalized medicine and an increasing occurrence of the development of a device to enhance therapeutic efficacy, the FDA has deemed it necessary to regulate such devices to assure physicians and consumers of the reliability of the analytical performance of the test.

7.2 DEFINITION OF AN IVD COMPANION DIAGNOSTIC

The first goal of the guidance document is to establish the definition of an *in vitro* diagnostic (IVD) companion diagnostic, in order to clarify further discussion. The guidance document describes an IVD companion diagnostic device as 'an *in vitro* device that provides information that is essential for the safe and effective use of a corresponding therapeutic product' [1]. The word *essential* here could potentially cause some confusion. Although those devices enhance the performance of a therapeutic, they are *not* required in order to gain approval for the therapeutic. To clarify this further, the document specifies those devices that:

1. identify patients that are most likely to benefit from a particular therapeutic product;
2. identify patients likely to be at an increased risk for serious adverse reactions as a result of treatment with a particular therapeutic product;
3. monitor response to treatment for the purpose of adjusting treatment (e.g., schedule, dose, discontinuation) to achieve improved safety or effectiveness [1].

While this definition would appear to clearly identify those devices intended to improve patient response to a particular therapy, there is some ambiguity in the statement 'clinical laboratory tests intended to provide information that is useful to the physician regarding the use of a therapeutic product, but that are not a determining factor in the safe and effective use of the product.' Given the definition provided in this guidance document, it would appear that there is still a place for laboratory developed tests (LDTs)[1] in association with a therapeutic product, assuming the product can achieve approval without the device. An example of this would be a device that may improve patient response rates in a subgroup, but in which the therapeutic response rates in the overall population are sufficient for approval of the therapeutic. While this has not been a frequent occurrence in the companion diagnostic devices that have been approved to date, it is important to note this nuance in the definition, as the regulatory approval process for an IVD companion diagnostic and laboratory developed test differ. An additional item of note is that this definition specifically[2] explains the history of the test and its relative impact on the device's classification.

---

[1] Typically well described tests used to diagnose diseases and conditions and historically intended to be used by physicians within a single institution. The components of traditional LDTs were regulated individually by the FDA as general reagents (though this is not the case entirely today), and the tests are developed and offered in CLIA-certified laboratories.

[2] FDA/CDRH Public Meeting: Oversight of Laboratory Developed Tests (LDTs), Date July 19–20, 2010.

Essentially, if the device is being utilized in the manner defined above as a companion diagnostic, then the history, or previous use and development of that device is irrelevant and the device would be subject to the same regulatory oversight as any other device. As the guidance states:

> 'An IVD companion diagnostic device that supports the safe and effective use of a particular therapeutic may be a novel IVD device (i.e., a new test for a new analyte), a new version of an existing device developed by a different manufacturer, or an existing device that has already been approved or cleared for another purpose [1].'

7.3 A MINI REVIEW OF FDA GUIDANCE ON APPROVAL OF THE COMPANION DIAGNOSTIC

The guidance document breaks down the approval process of the companion diagnostic into two major classifications – novel therapeutics and approval of a therapeutic product without an approved IVD device. For novel therapeutic products, the intention is for the companion diagnostic to be developed and approved in conjunction with the novel therapeutic. This intention is evident based on the previous description of an IVD companion diagnostic. As the companion diagnostic is required for safe and effective therapeutic use, the approval of the companion diagnostic would be therefore required for the therapeutic approval. The second classification provides more complicated scenarios. In these cases, a therapeutic product may have been previously approved or the FDA has determined that it is appropriate to approve the therapeutic without the benefit of the companion diagnostic device. While the definition of an IVD companion diagnostic indicates that the device is required for approval, the guidance clarifies the scenarios in which a therapeutic could be approved without prior or concurrent approval of the diagnostic device. The two scenarios in which this may apply are if the therapeutic is a novel approach for the treatment of a serious or life-threatening condition, or if the therapeutic is already approved. In the first case, the benefits of the potential treatment must outweigh the potential implications of not having the approved device available. In the second case, the FDA will not generally revise a label without the companion diagnostic approval, but in cases where serious safety issues are at play. An example of this case would be the revision of the labels of the epidermal growth factor receptor (EGFR) inhibitor cetuximab to include testing for KRAS mutation in codons 12 and 13 despite there being no approved diagnostic device, based on strong clinical data [4]. Despite these two scenarios, it is generally assumed that approval of the diagnostic device will be required for approval of the therapeutic, and therefore the sponsor will appropriately plan for the device approval in the course of their clinical design. In addition to this general policy, the FDA has laid out a number of additional general policies. These policies include a risk-based approach to assignment of a regulatory pathway; the approval of a therapeutic and its companion diagnostic are intended to be reviewed and approved in concert. Should a pre-approved device be utilized for a novel therapeutic device, it will be reassessed under a new premarket approval (PMA) filing, and a new IVD companion diagnostic for a previously approved use will also be reviewed under a separate submission.

7.3.1 Labeling

The labeling of both the therapeutic and diagnostic are also addressed in the draft guidance. Existing regulations dictate that:

'for drugs and biological products (21 CFR 201.56 and 57), product labeling must include information about:

1) specific tests necessary for selection or monitoring of patients who need a drug;
2) dosage modifications in special patient populations (e.g., in groups defined by genetic characteristics); and
3) the identity of any laboratory test(s) helpful in following a patient's response or in identifying possible adverse reactions [1].'

The previously issued regulations also address sections in which to include the appropriate language. The guidance provides clarification in that the novel therapeutic will include information in its label about the co-developed companion diagnostic device. When appropriate, the label should indicate the type of approved or cleared companion IVD companion diagnostic device, and in cases where the IVD companion diagnostic device is approved after the therapeutic, the therapeutic label should be updated to reflect the newly approved IVD companion diagnostic. Regarding device labeling, the guidance reflects that the intended use of the diagnostic should be included in the labeling for the device. Any subsequently identified uses, for example in another indication, or for another therapeutic, would require a new premarket submission to amend the labeling of the diagnostic device to reflect that additional use.

7.3.2 Investigational Use

While the primary focus of the guidance document is on the filing of and approval for IVD companion diagnostics in association with a therapeutic, the draft guidance also covers the use of devices in the scope of clinical trials. As such investigational devices are defined as:

'All diagnostic devices used to make treatment decisions in a clinical trial of a therapeutic product will be considered investigational devices, unless employed for an intended use for which the device is already approved or cleared [1].'

In addition to this definition, there is clear language stating that if a device is used to:

'make critical treatment decisions, such as patient selection, treatment assignment, or treatment arm, a diagnostic device generally will be considered a significant risk device under 21 CFR 812.3(m)(3) because it presents a potential for serious risk to the health, safety, or welfare of the subject, and the sponsor of the diagnostic device will be required to comply with the investigational device exemption (IDE) regulations that address significant risk devices [1].'

This means that the device must follow IDE regulations throughout its use in the clinical trial, and that any combination of a diagnostic and therapeutic trial must comply with both IDE and Investigational New Drug (IND) regulations. In order to comply with the IDE regulations, the FDA suggests that information about the planned intended use of the device be shared with the appropriate product review center, and also strongly suggests discussing the IVD companion diagnostic device via pre-IDE submission. It is important to note these suggested interactions, as these meetings must be appropriately planned for and timed in accordance with the proposed filing time line for both the therapeutic and the device.

7.3.3 Summary

The regulatory pathway for the IVD companion diagnostic device has historically not been well defined, and varies widely depending on the governing regulatory authority. Recently, regulatory guidance such as that provided in the Draft Guidance for Industry and Food and Drug Administration Staff on *In vitro* Companion Diagnostic Devices that we reviewed here help to clarify this ambiguity [1]. However, the guidance provided is only a draft, is described as representing the current thinking of the FDA, and is not binding. With this in mind, it is important to have frequent and candid interactions with the regulatory authority overseeing the device for your region of interest to ensure that opinion and guidance does not change or differ from what has been provided publicly. In addition, each diagnostic device has particular technical aspects which may alter the regulatory approach, and therefore accurate guidance specific to the device should be sought for each new device, therapeutic indication, or device application.

7.3.4 Timing Considerations

As depicted in the previous review of the FDA guidance on *In vitro* Companion Diagnostic Devices, the FDA strongly suggest a number of different interactions with regulatory authorities throughout the assay development and implementation process. Figure 7.1 is taken from the draft April 2005 Drug Diagnostic Co-Development Concept Paper [3]:

There are two considerations for the proper timing of obtaining CDRH feedback on the device and therapeutic combination. The first consideration is the appropriate timing to provide ample data for CDRH to review and provide feedback. In other words, when do we have enough information to solicit feedback? The second component regards the timing associated with approval of the diagnostic or gathering feedback appropriate for implementation in a certain clinical phase. The first consideration is rather straightforward, but an important distinction. The FDA can provide limited feedback on mere plans, but may be able to give more insight with data in hand. This consideration is important when planning for analytical and clinical validation. For example, while validation plans may be reviewed by the FDA via the pre-IDE, without some data regarding assay qualification and establishment of performance characteristics, the reviewers would have a limited frame of reference to determine whether validation plans and numbers are sufficient for a particular assay. If not properly planned for, validation may need to proceed at risk. The second timing consideration is much more complex, as the time lines for approval of the therapeutic will often change over the course of the product's development. While there are many factors in play, the ultimate goal is to have the filing of the device precede or be in concert with the filing of the therapeutic. In order to effectively accomplish this task, the commercially ready test should be employed in pivotal phase testing, and all interactions necessary for its implementation should be completed prior to pivotal phase start.

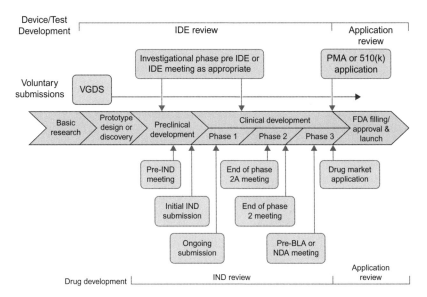

FIGURE 7.1   The graphic presented here from the Drug Diagnostic Co-Development Concept Paper depicts the ideal timing of regulatory interactions and filings in concert with therapeutic filings.

7.3.5 Regulatory Pathways Outside of the United States

There are a number of regulatory considerations to consider when a companion diagnostic device is registered outside of the United States. As mentioned previously, it is currently believed that the US device approval is the most stringent of the regulatory paths and even considered by some to be a hindrance to advancing the field [5,6]. However, other regulatory bodies also have stringent approval processes in place. For example, the Japanese Pharmaceutical and Medical Devices Agency (PMDA) have specific guidance indicating the necessity for foreign manufacturers to have a marketing authorization holder in Japan to sponsor the device and oversee the approval process [7]. These requirements must be carefully considered and strategically planned for to ensure appropriate timing of medical device filings in multiple geographic locations. In addition to the PMDA requirements, additional regulatory bodies have established guidelines and general provisions for the review and approval of medical devices. An example of this is the 'Regulations for the Supervision and Administration of Medical Devices', released by the Chinese State Food and Drug Administration (SFDA) [8]. While it will not be covered in detail here, the document lays out clear requirements and penalties for not fulfilling those device requirements. In contrast to Japan and China, where there is clear oversight of device approval by a regulatory authority, Europe utilizes a CE marking system which applies to all devices, not only medical devices. The European Commission has issued a directive for the required documentation required of the sponsor of *in vitro* diagnostic devices, yet specific regulatory approval has not been required [9]. Instead the sponsor is required to prepare documentation stating that the analytical and clinical performance of the device is reliable [9]. While this is only a cursory glance at the regulatory pathway to commercial use of an *in vitro* diagnostic in association with a therapeutic agent, it is meant to present the complexity of approval of an *in vitro* diagnostic in multiple geographic locations under the authority of a number of regulatory bodies. It is critical to understand the requirements of the specific geographic location in which you seek approval or clearance of the diagnostic to ensure appropriate planning and design of analytical and clinical validation studies.

7.3.6 510(k) vs PMA

There are two types of device approvals and clearances. The distinction for a device is made as either Premarketing Notification or a 510(k), or a Premarket Approval or PMA. The key distinction from the FDA viewpoint is the level of risk of the device. A 510(k) is for lower risk devices where the FDA requires only substantial equivalence to a currently marketed device for which a PMA is not required [10]. The FDA defines *substantially equivalent* in the following manner:

- has the same intended use as the predicate; and
- has the same technological characteristics as the predicate; or
- has the same intended use as the predicate; and
- has different technological characteristics and the information has been submitted to the FDA;
  - does not raise new questions of safety and effectiveness; and
  - demonstrates that the device is at least as safe and effective as the legally marketed device [10].

The FDA may require different information from different sponsors depending on the intended use of the device, the technology involved, the therapeutic area to which it is being applied, etc., so there is no uniform data package to ensure demonstration of substantial equivalence. The ultimate decision from the FDA on a 510(k) submission is a clearance decision, not an approval decision. The 510(k) definition is important to understand in the diagnostic development process if you are developing a device similar to a currently marketed device. The FDA has in numerous public meetings and also has specified in the most recent draft guidance [1] that devices intended to inform a clinical decision on a therapeutic are considered high-risk devices and are therefore subject to the PMA process. The PMA 'is the most stringent type of device marketing application required by the FDA [11].' Rather than a clearance decision, the PMA is either granted or refused approval. As the FDA states:

> 'PMA approval is to be based on a determination by FDA that the PMA contains sufficient valid scientific evidence that provides reasonable assurance that the device is safe and effective for its intended use or uses [11].'

Due to this distinction, much more thorough data are required in the PMA submission. The burden for PMA approval stated above should be strongly considered when developing, validating, and applying a companion diagnostic with a therapeutic product.

7.4 ASSAY DEVELOPMENT AND VALIDATION IN DRUG DIAGNOSTIC CO-DEVELOPMENT

7.4.1 Overview

The process and evolution of an assay from its initial discovery to a form with potential clinical applicability is difficult to generalize, as there are many technical aspects that may dictate deviation from the standard process. The draft Drug Diagnostic Co-Development Concept Paper of April 2005 [3] presents the following ideal assay development progression (Fig. 7.2):

The timing, target selection, and validation presented here are very pre-clinically driven, which is not likely feasible as validation and hypothesis generation require a significant amount of patient data. However, with the diversity of biomarker analytes and potential applications, there is a general evolution that most assays encounter, and therefore we will approach the validation of an assay using the general evolution as a standard flow. For our purposes, we will define the evolution of a companion diagnostic assay as falling into the following categories:

1. Research Assay – the assay in its initial discovery stage
   a. Platform and assay layout may not be ideal for clinical application
   b. Assay components are often research use only
   c. Assays have not been validated, often still in 'off the shelf' format
   d. Applied to early pre-clinical and early-stage clinical development samples in hypothesis generation and testing phase

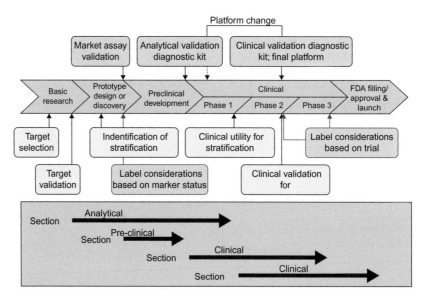

FIGURE 7.2   This figure, taken from the Drug Diagnostic Co-Development Concept Paper, represents an ideal timing scenario and level of validation for an assay as it progresses through development into a companion diagnostic in concert with therapeutic development time lines.

2. Clinical Trial Assay – the first stage in migration to clinical applicability
   a. Assay migrated to platform and format applicable to clinical testing environment
   b. Assay components migrated to as many good manufacturing process (GMP) grade reagents as possible, with avoidance of assay-specific reagents
   c. Assays involved are minimally qualified to establish early performance criteria
   d. Applied to later stage clinical development as an investigational use device
3. Investigational Usage Only (IUO) Assay – assay further refined for clinical implementation, represents the final evolution prior to clinically relevant device
   a. Assay is applied to final commercially available platform in appropriate format
   b. Assay components are all GMP grade with no assay-specific reagents
   c. Assays involved are fully validated and evaluated against previously established performance criteria
   d. Applied in pivotal stage clinical testing to prove hypothesis and provide clinical validation
4. IVD – the final commercially available cleared device
   a. Clinically validated device applied to therapeutic product available for physician use and indicated in therapeutic label.

A graphical representation of this progression is given in Fig. 7.3. While assays may be in a hybrid of the stages described here, this general evolution of an assay is followed by the majority of companion diagnostic devices. As such, it is important to note the stage of the assay you are developing and make sure it is being applied in the correct stage of development with the correct level of analytical and clinical validation. These considerations will be discussed in greater detail throughout the remainder of the chapter.

7.4.2 General Considerations in Drug Diagnostic Co-Development

Prior to discussing the level of validation and considerations for implementation of potential diagnostics in clinical development, there are some general concepts of drug diagnostic

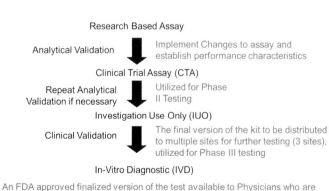

FIGURE 7.3 A graphical representation of the natural progression of an assay from discovery through companion diagnostic with levels of validation necessary included for each step.

co-development that should be addressed. The first of these concepts is the implementation and utilization of the assay in clinical trial design.

7.4.2.1 *Implementation of Companion Diagnostic in Clinical Design*

To simplify the possibilities of how a diagnostic can be implemented in the clinical trial setting, we will explain the three potential designs as noted in the draft Drug Diagnostic Co-Development Concept Paper [3], and discuss the benefits and drawbacks of each of the proposed designs. The three designs are the classic two-arm clinical design, randomization with the diagnostic, and the restricted approach. The classic two-arm clinical design is depicted in Fig. 7.4.

In this approach all subjects that meet enrollment criteria are randomized into either drug or placebo arms regardless of their diagnostic status. All subjects are still evaluated at baseline for the diagnostic status, but the diagnostic status is not utilized in the trial design. This is the most classic of the design types and allows only for retrospective analysis and does not allow for adaptive clinical trial design. In addition to only allowing a post-hoc analysis approach, this strategy also opens up the investigator to the potential of enrolling the majority of the signature positive or negative patients into the placebo arm. Without prior understanding of the prevalence of the diagnostic positive and negative patient population, this strategy would not be the preferred approach for this reason. While there are significant drawbacks in this approach, there are also some benefits. For example, classical trial design does not slow enrollment by waiting on the results of a diagnostic test. This will likely result in more rapid patient enrollment, albeit at the risk of biased and unbalanced stratification by diagnostic status in the treatment and placebo arms. In addition to the likely benefit of shortened enrollment, this design also allows for collection of samples at baseline to be tested later with the diagnostic assay, assuming stability of the analytes. This benefit is most readily applicable when an assay has not reached the appropriate level of validation prior to trial initiation. The design would allow time for validation of the assay to the appropriate level prior to the testing of clinical samples. Finally, this trial design allows the investigator to get a sense of the prevalence of diagnostic negative and positive subjects to be applied to future clinical development. This trial design is most commonly applied early in clinical development, and does not result in a significant confirmation of hypothesis testing due to the potential for positive and negative subjects to be unevenly distributed in the treatment and placebo arms. It is therefore not a recommended approach when seeking to test a diagnostic hypothesis.

The second approach is to randomize the trial by diagnostic status. A diagram of this approach from the draft Drug Diagnostic Co-Development Concept Paper [3] is shown

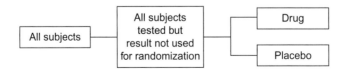

FIGURE 7.4   A figure taken from the Drug Diagnostic Co-Development Concept Paper depicting a classical two-arm clinical design in which patients are randomized into drug or placebo without regard for the diagnostic call.

in Fig. 7.5. In this design, all subjects are tested for diagnostic status at baseline and then randomized into a drug or placebo cohort based on the diagnostic test result. The drawback of this trial design is that it requires the test to be available at trial start for implementation. The test must also provide rapid results to ensure timely enrollment and stratification. In addition, depending on the distribution of negative and positive subjects at screening, it may slow enrollment if the investigator desires equal distribution in the test positive and negative groups. In terms of benefits, the trial design allows for the prospective identification and testing of the diagnostic hypothesis. The patients are equally distributed among test negative and positive groups and drug and placebo arms, giving the clearest picture of diagnostic predictive value. True positive and negative predictive values can be calculated from this clinical trial design, as clinical response to therapeutic intervention will be available for both diagnostic negative and positive subjects. This approach also allows the investigator to understand the performance of the diagnostic in a more commercial setting with result turnaround times intended to mimic that required by a physician in practice. Based on the benefits outlined above, this approach is the preferred clinical trial design in the evaluation of diagnostic hypotheses. This approach is often used in early and later stage clinical development where there is a better understanding of diagnostic prevalence and potential predictive value.

The final clinical trial design for diagnostic assay implementation is the restricted approach. In this design the trial enrolls only those patients that are diagnostic positive. A diagram of this clinical trial design is depicted in Fig. 7.6.

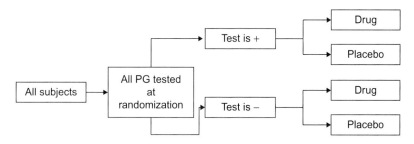

FIGURE 7.5 A diagram of the randomization by diagnostic call clinical design as depicted in the Drug Diagnostic Co-Development Concept Paper.

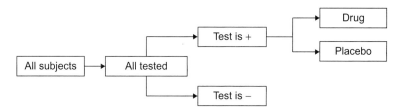

FIGURE 7.6 A graphical representation of the restricted clinical design approach as shown in the Drug Diagnostic Co-Development Concept Paper. Represents an aggressive clinical trial design in which only diagnostic positive or negative patients are enrolled in a trial.

This is a very aggressive clinical trial design in which one must have strong confidence in the scientific link between diagnostic status and drug response. This design has a number of significant drawbacks. Perhaps the most critical one is the lack of understanding of the response rates in the diagnostic negative patient population. Excluding patients from treatment without understanding the potential clinical response rate is a high-risk strategy and one not easily accepted by the FDA. As a result, this strategy is predominantly utilized only when the diagnostic is predicting some significant toxic response, and by preventing these subjects from receiving the therapy the benefit-risk profile of the therapeutic is increased. Should this be the case, the therapeutic would certainly be limited in its label to include only test-positive patients. Another drawback of this approach is that the test would have to be nearly fully validated to be utilized in this manner. This limits the use of this design to later stages of development only, as data to properly validate the test as well as to indicate the toxic response occurring only in the negative group would be required from early phase trials. This would require a significant number of subjects to demonstrate statistical significance; most likely a very large Phase II trial is needed to show that the therapeutic of interest does not benefit patients in the diagnostic negative group and is statistically different from that of the diagnostic positive group. In terms of benefit, this trial design is likely to enroll quickly if the prevalence of the diagnostic positive subjects in the population is high. Usually, this is not the preferred design of either the FDA or the investigator. Due to the risk and aggressive nature of this design, it is employed only in those cases where there is a direct link between the diagnostic and drug mechanism of action (MoA) or where there are significant safety concerns in the diagnostic negative patient population.

7.4.2.2 *Establishing a Training and Test Set of Samples*

The process of training and testing as two independent steps in the development of a biomarker classifier is very important, though sometimes incorrectly applied. A training set, by definition, is the data set that is used to optimize the features (e.g., transcripts, proteins, cellular subtypes), account for the variability within the target population, and ultimately establish a cut-point to discriminate conditions or groups. A test set, also called a validation set or hold-out set, is an independent data set with similar characteristics to the training set, used to evaluate the performance of the classifier, where *performance* can be defined by metrics such as sensitivity, specificity, predictive value positive/negative, and/or area under the curve [12,13]. There are methods to utilize the training set and perform cross validation procedures to mimic the structure of a test set. These approaches can range from *n*-fold random or stratified splits to leave-one-out cross validation, though it can be argued that true performance evaluation is best represented using a completely independent data set for testing, and as such, the discussion provided here will focus on the training/test set design.

For appropriate use of a test set, there should be no structure from the training set or training process introduced. The test set should serve as a *real world* application of the classifier performance, such that all of the model parameters that have been optimized in the training process, should be held constant and not modified once model testing has commenced [14]. Though this concept appears straightforward at first glance, there are various points of either disparity or contamination between the two data sets that can be overlooked, thus introducing bias leading to an overfit result into the biomarker classifier development process. The next few sections discuss some of the more prevalent inconsistencies or potential pitfalls in classifier development for a biomarker application.

7.4.2.3 *Patient Population Consistency*

One of the most important principles in identifying and developing a biomarker classifier is characterization and subsequent selection of the target patient population. Consistency in the population should be maintained in both the training and test sets by matching on known relevant patient baseline variables. Multiple human genetics studies have demonstrated the inherent differences between race, ethnicity, and/or gender and the ways the modes within these patient variables can differ with respect to biological processes such as gene/protein expression patterns or clinical outcomes. Further, patient characteristics or lifestyle variables such as comorbidities, medications, smoking status, and alcohol consumption status can all potentially drive observed biological alterations. Disease heterogeneity is another key factor that should be controlled for, particularly in indications with known subtypes of disease. All of these potential sources of disparity in the patient population should be accurately assessed in the training set and maintained in the test set to ensure consistency in the biomarker performance.

7.4.2.4 *Iterative Evaluation Cycles*

A potential source of bias in the biomarker development process between the training and test set is the concept of multiplicity, or excessive refinement iteration. This can occur if the training and test sets are utilized simultaneously in an iterative model optimization progression. For example, if parameters are optimized on a training set and then applied to the test set to evaluate classifier performance, and this cycle is repeated multiple times until an optimal result is obtained, there is a concern for overfitting, or non-generalizability of the classifier [15,16]. Although, for this example, the test set was not utilized in the parameter optimization or training process, the iterative fitting of the training set with multiple evaluations using the test set can be viewed as a biased approach. Depending on the number of cycles conducted using this approach, as well as the number of patients in each data set, it is possible to reach convergence to an acceptable performance, and likely overfit performance accuracy, simply by the sheer number of iterations. By using this practice, typically the training and test set classifier performance metrics are evaluated side by side within each iteration and the ultimate model parameters are selected on the basis of either the top ranking test set prediction result or, more likely, one of the top ranking test set prediction results and prediction concordance with the training set performance. Both of these selection scenarios can lead to an overfit result and the fundamental difference between these approaches, as compared to an iterative fitting process on the training set *only* (with minimal test set predictions conducted), should be understood.

7.4.2.5 *Clinical Material Considerations*

Regardless of the clinical design employed, the acquisition of sufficient clinical material to analytically validate the diagnostic assay is of utmost importance. Frequently, throughout the history of an assay's migration to companion diagnostic status, a number of changes will occur to the assay to enable a more commercially viable test. In the course of these changes, equivalence in call to previous versions of the test must be established. This is known as bridging. Bridging is the testing strategy employed whereby a sample is tested with two different versions of the test to demonstrate equivalence in diagnostic call. Optimally, the samples utilized to bridge the assay would be those samples comprising the training and test sets. This allows for accurate setting of the classifier in each iteration of the test. Each time a test is bridged from one version to another, amounts of sample are consumed in the process. A number of bridges of an assay can quickly utilize all of the available samples. This point cannot be understated, as sample availability can be a limiting factor in the ability to optimize an assay. If all sample material has been exhausted, the accuracy of the bridge can be brought into question, or additional samples may be needed to strengthen the migration to the updated version of the assay. Given the importance of sample availability, it is key to collect sufficient samples from early clinical development programs to support the development of the diagnostic. This often includes collecting as many aliquots from each patient as possible. Because collecting as many samples from each patient is of such importance, the clinical material utilized in the assay must be readily available. Therefore, it is important to ensure either correlation between disease site information and a surrogate tissue, or sufficient access to the disease site tissue itself. However, often obtaining multiple biopsies to access disease site tissue samples is not a safe or logistically feasible option. In most indications, this means developing assays in more readily available biofluids such as whole blood, serum, or plasma, or utilizing paraffin embedded tissues as starting material for diagnostic assays after establishing correlation with the diseased tissue. Utilization of these more readily available tissues helps improve the potential for having sufficient clinical material for assay development and validation, and also eases the transition to commercial clinical testing laboratory implementation. A diagnostic requiring disease site tissue to function properly will likely be limited not only by sample for assay development, but also availability of sample from patients in the commercial clinical testing setting. Many patients may elect not to have the sample taken unless the test has an extremely powerful predictive value.

7.4.2.6 Technical Considerations

In addition to the aforementioned general considerations for diagnostic development, there are also a number of technical considerations that must be addressed to ensure successful translation of a diagnostic from discovery through clinical validation.

According to feedback obtained by interaction via pre-IDE meetings with the FDA, it has been noted that most changes to an assay are considered major or substantive changes and therefore repeat validation following these changes would be necessary. This includes, but is not limited to, enzymes, buffers, software, algorithm, and instrument changes. With this definition in mind, the companion diagnostic developer must be aware of the status of each component of the assay.

7.4.2.7 Testing Volume

Extensive marketing research is required to determine the expected volume of testing for a particular companion diagnostic device. The components of this include the therapeutic disease prevalence, physician-required turnaround time, current standard of care treatment, competition in the market, availability of platform in testing market, and configuration of the assay itself, as well as additional factors. It is important to conduct this research early in the development of the companion diagnostic assay, as feedback from this research can guide the development of a more market-ready assay. For example, if the test will be relatively low volume due to low disease prevalence and a less time-sensitive turnaround, then multiplexing assays to optimize the number of samples that can be run on a single plate may not be necessary. However, if the disease of interest is highly prevalent, and research indicates that the majority of patients will be tested, then the assay format would need to be adjusted to optimize the number of samples that can be evaluated on each run. The volume of testing may also dictate whether the test is available in all clinical testing laboratories throughout the region of interest, or if specific laboratories are targeted to run the companion diagnostic device. It is wise to have multiple parties conduct similar research in this effort to ensure an accurate assessment of test volume, as it has been noted that test volume alone can be a major contributor to assay format and development.

7.4.2.8 Ease of Use

Regardless of the number of sites necessary to handle testing volume, the ease of use and result is a key to a successful companion diagnostic implementation. Complex assays that require a great deal of manipulation increase the possibility of user error. For this reason it is imperative that the device, including everything from analyte isolation through to final result, be as user friendly as possible. There are a number of strategies to improve ease of use. These strategies generally fall into one of two categories – automation or assay simplification. While automation may be considered a viable option for a number of devices, the investigator must remember to consider the clearance status of the automation equipment being utilized. There are only a few pieces of automation instrumentation and associated protocols that have been cleared. The advantage of automation is that it significantly decreases the opportunity for user error leading to test failure. Automation is best applied to repetitive tasks, freeing technicians to perform other steps requiring more user input. Automation can be applied throughout the process from analyte isolation through plate or assay set up, and perhaps is best applied in terms of analysis. Software including automated pass/fail, and subsequent generation of results, removes the user from performing any calculations and provides consistency to the analysis. Assay simplification is another key component of ease of use. This includes aspects of the device process, such as multiplexing for multiple targets, combining steps if possible (i.e., one step polymerase chain reaction (PCR) instead of a separate reverse transcription reaction and quantitative PCR steps), and minimizing the number of pipetting steps. While alterations to the assay such as these may seem trivial, they must be incorporated early in the development of an assay to ensure there are no additional technical hurdles created. In addition, later incorporation of these changes may require additional validation work that will cost both time and resources. The investigator needs to be aware of the risk and benefit of the addition of these measures in the assay processes and decide which aspects are necessary for the particular assay in question. In general, the more simple and user friendly an assay can be, the better, as this will ease the transition of the device into the clinical testing setting.

7.4.2.9 Number of Analytes

The number of analytes assayed can be quite relevant to the development of a companion diagnostic device. From a validation standpoint, draft guidance has suggested that a test featuring two to 10 analytes must be individually validated, while larger numbers of analytes dictate validation of the system as a whole rather than the individual components [3]. There are also considerations of assay configuration based on proposed sample throughput. As discussed in the testing volume section above, a large number of analytes may limit the numbers of subject samples that can be evaluated in a single run. If throughput is an important factor to the device, then higher numbers of analytes may bring the feasible use of the diagnostic into question. In addition, the numbers of analytes in an assay often determine the price of an assay. This is due to the code stacking method of reimbursement applied to newly approved devices. Manufacturers of the device typically will take existing current procedural terminology (CPT) codes and use them multiple times to account for each reaction that is taking place. The use of putting the same codes down to an account multiple times for the total cost of a procedure is termed code stacking. As described in a recent *American Association of Clinical Chemistry* article:

> 'Unlike traditional assays, molecular tests do not have single, analyte-specific codes that labs can use to bill Medicare or private payers. Rather, a list of codes that signify each procedure involved in performing the assay are listed together. Each code describes a separate step or methodology performed to complete the test, such as gene amplification, nucleic acid extraction, or nucleic acid probes. For some tests, these codes must also be multiplied if the laboratory uses a step more than once to perform the test, especially in those tests that look at multiple markers. As an example, if a PCR-based assay evaluates seven targets and three house-keeping genes to determine a classifier, then codes for molecular extraction, nucleic acid transfer, and nucleic acid PCR at a minimum would be stacked for each of the 10 reactions per sample. Changes to the coding scheme have been made recently that are aimed at ending code stacking and separating molecular tests into two major tiers, adding them to the physician fee schedule and away from the clinical laboratory fee schedule. However, implementation of this coding method has yet to be fully adopted due to a number of complex issues surrounding reduced reimbursement for clinical laboratories [17].'

Regardless of the coding method being employed, it is important to understand the interplay between the number of analytes in a device and cost of the device.

7.4.2.10 *Multiple Analyte Versus Few Analyte Classifiers*

A very common concern in biomarker classifier development is the large number of analytes measured for a given patient. This issue has become almost commonplace in the last 15 years with the introduction of microarrays used to measure thousands of transcripts, to multiplex protein and cellular assays, to the more recent use of high-throughput sequencing. The number of analytes or variables can far exceed millions for a given patient specimen. This type of situation is known as the 'curse of dimensionality', where an estimator will converge to the true value of a smooth function on a high-dimensional space very slowly [18,19]. Or more specific to a biological context, this expression means that in order to obtain a good estimate for a function of analytes to indicate a predefined expression pattern, a very large number of patient specimens would be required [20]. However, the ability to procure such large numbers of specimens is often hindered by either cost or availability. There are factors that can often reduce the initial set of analytes to a smaller set of useable components for a classifier, and various statistical methods have been developed to summarize multiple analytes into a low-dimensional function that can be utilized to elucidate meaningful results.

For practical implementation, fewer analytes in a classifier can be argued as preferable to a large panel of analytes. This concept follows from the issue of dimensionality explained previously, where if one were to attempt to partition p patient specimens into two groups (e.g., tumor vs. normal tissue) based on g genes, where $g >> p$, it is very likely that a classifier based on some set of genes would be capable of doing so into the appropriate two groups with 100% accuracy. This result, however, would probably not be generalized enough to perform with the same accuracy on an independent set of patients. Further, among those genes in the function, there would exist a large amount of mutual or correlated information, thereby providing redundancy in explaining the separation between the two patient populations. To address this issue, statistical methods have been developed to remove those variables that do not significantly contribute to the decision rule or classifier and are highly correlated with other variables [21]. Often, these types of methods can greatly reduce the analytes in the model to the most predictive set and still maintain a very similar prediction performance. Various error functions have been used to determine this cost/benefit ratio in the optimization of the best predicting variable set.

7.4.2.11 *Instrument*

In order to ease transition and prevent additional validation work, it is recommended to utilize instrumentation and associated software that has already been through the clearance process and has obtained 510k clearance. This will reduce the regulatory hurdles encountered in a first-time submission of a platform and associated software. This can be either a trivial decision, or much more complex, as most commercial diagnostic development partners are closely tied to development on equipment for which they own development rights. For example, Qiagen develops assays on their proprietary platform, the Rotor-Gene Q[3]. Early development on the ultimate clinically applicable platform can offer significant advantages such as decreased bridging, early understanding of performance characteristics, and early optimization of assay format. If it is possible to understand the ultimate commercial partner for the assay early in assay development, the appropriate platform can be utilized to avoid assay transfer and migration to a different platform late in the development process. In addition to the regulatory and sample requirement considerations of the assay platform, the investigator must also consider the footprint of the instrument utilized in the device. For example, if a diagnostic assay is built on an instrument which is not present in the geographically necessary clinical testing laboratories, there will be additional delays to test distribution as these laboratories evaluate the risks and benefits of procuring the new platform. With this in mind, it is wise to understand the footprint necessary for rapid adoption of the particular companion diagnostic assay of interest and ensure that this is either already present or appropriately planned to meet that footprint in advance of device approval.

7.4.2.12 *Reagent Status*

In the pre-IDE setting, the FDA closely evaluates each of the specific reagents included in the assay. For each reagent, the manufacturing status must meet that of a commercial assay prior to approval. This means that each assay component must be manufactured to Good Manufacturing Practices standards. In the past, the FDA had allowed use of what was termed analyte-specific reagents, but only in the context of laboratory developed tests (LDTs), which we now know from guidance not to be an appropriate regulatory pathway for companion diagnostics. In addition to ensuring the appropriate level of manufacturing compliance, the investigator must identify critical reagents in the assay. For these critical reagents and components, the FDA looks for alternative manufacturers or risk mitigation strategies in the scenario where a manufacturer may discontinue the production of a certain component. Therefore,

---

[3]http://www.qiagen.com/products/rotor-geneq.aspx#Tabs=t1.

ideally the ultimate commercial distributor should appropriately control all aspects of the assay. If multiple vendors are used to source different portions of the assay, then risk mitigation and identification of alternative vendors can become quite complex.

7.4.3 Assay Validation

A key component of any assay is the ability to demonstrate that it provides reliable and accurate assessment of the analyte of interest. In order to demonstrate that this is true, the assay must by analytically validated. While the components of an analytical validation may vary depending on the type of assay, a systematic approach to validation is required to ensure that all necessary components are addressed. The analytical validation of a device can be broken into two distinct steps. First, the performance characteristics of an assay are established during assay characterization or qualification. This process allows for the identification of characteristics and standards that the assay must meet in the second step, formal analytical validation. The actual validation tests the assay in a realistic setting to assess the device's ability to meet the specifications laid out in the assay characterization.

7.4.3.1 Assay Characterization or Qualification

Assay qualification and characterization are keys to a successful device analytical validation, as this step lays the groundwork for the formal validation procedures. Often an assay is developed or utilized in a research setting during hypothesis generation and following the assessment of the predictive value of the proposed companion diagnostic in early trials. This logistical issue leads to an insufficient understanding of the assay's performance characteristics prior to its migration to a more clinically applicable version. In addition, sufficient changes to a research assay to accommodate a better clinical design necessitate that performance characteristics will need to be redefined. During assay performance characterization, the assay should be subjected to conditions outside of what is expected in the clinical setting. This testing at the extreme ends of expected conditions allows the investigator and assay developer to assess the breaking points of the assay and evaluate a reliable range of performance. To ensure appropriate criteria setting for all aspects of validation, each of these aspects should be thoroughly assessed. A complete list of parameters to assess is dependent upon each assay, and a more thorough description of the validation parameters is covered in the following section.

7.4.3.2 Analytical Validation of a Companion Diagnostic

The analytical validation of a companion diagnostic can be a complicated process, but is simplified utilizing a systematic approach to its completion. The aspects of an analytical validation are defined in the draft Drug Diagnostic Co-Development Concept Paper [3] as the following:

1. Studies to show that test performance can be applied to expected clinical use as a diagnostic with acceptable accuracy, precision, specificity and sensitivity: A demonstration of the device's ability to accurately and reproducibly detect the analyte(s) of interest at levels that challenge the analyte concentration specifications of the device should be provided (see point 3 below).

2. Sample requirements: All relevant criteria and information on sampling collection, processing, handling and storage should be clearly outlined.

3. Analyte concentration specifications: It is recommended that, when appropriate, a range of analyte concentrations that are measurable, detectable, or testable be established for the assay.

4. Cut-off: It is recommended that there be a clear rationale to support an analytical characterization of cut-off(s) value(s).

5. Controls and calibrators: All external and process controls and calibrators should be clearly described and performance defined.

6. Precision (repeatability/reproducibility): All relevant sources of imprecision should be identified and performance characteristics described.

7. Analytical specificity (interference and cross reactivity studies): Cross-reactive and interfering substances should be identified and their effect on performance characterized.

8. Assay conditions: The reaction conditions (e.g., hybridization, thermal cycling conditions), concentration of reactants, and control of nonspecific activity should be clearly stated and verified.

9. Sample carryover: The potential for sample carryover and instructions in labeling for preventing carryover should be provided.

10. Limiting factors of the device should be described, such as when the device does not measure all possible analyte variations, or when the range of variations is unknown.

While accurate descriptions of the methods required for proper analytical validation of a drug or biologic are clearly defined in FDA guidance [22], there are not clear steps for the device navigation process, and as such, the topic will be addressed here. For the purpose of description, we will treat the criteria listed here as a checklist of tasks to be completed, and describe each step in detail in the following sections.

7.4.3.3 *Studies to Demonstrate Accuracy, Precision, Specificity, and Sensitivity*

Appropriate studies must be designed to accurately assess the reliability and reproducibility of a device. This assessment requires two critical components. The first of these is the production of a synthetic target to assess the ability of the assay to detect the specific analytes involved. For example, in the case of gene expression-based diagnostic devices, this surrogate for the analyte of interest in a biological specimen is an *in vitro* transcribed synthetic target generated in a plasmid. These surrogates serve a number of purposes in validation. The surrogate can assess that ability of a particular assay for its specificity to a particular analyte. Also, the surrogate can assess the limits of quantitation and detection of each assay by accurate dilution to known input concentrations. Often times these synthetic analytes can also be utilized as run controls for the final product release. The second component necessary for successful determination of accuracy and precision is the analytical reference method. The analytical reference method is a previously defined method capable of assessment of the analytes of the assay in a similar or superiorly sensitive fashion yet utilizing different biological and chemical methods. Often the analytical reference method is referred to as the gold standard. As an example, a gene expression-based microarray companion diagnostic would utilize qPCR as an analytical reference method as it provides similar data from an equally or more sensitive chemistry. This allows for the assessment of accuracy as required in this section. Not all methods will have an easily identified analytical reference method. In these cases it becomes quite difficult to assess the accuracy of a diagnostic device when there is no true answer against which to compare results. While assay performance characteristics of individual assays can be established utilizing controls, it is important that the validation itself is carried out with patient samples identical to the intend-to-treat population if at all possible. This will be practical data that reviewers can utilize to assess the performance of the device in a real world setting. Analytical studies utilized here include specificity, accuracy, precision, and robustness.

7.4.3.4 *Sample Requirements*

There must be clear criteria laid out for the collection and storage of all sample types utilized for the device. This includes collection methodology, stabilization methods, appropriate storage conditions, and stability of the analytes at the stated storage. Ideally a method will be employed that has been previously cleared. In utilizing a previously cleared sample collection and extraction product, the storage conditions will have been previously defined. Also of note on this aspect of analytical validation is that the sample type should be stable in order to accommodate the test. At a minimum, samples will have to be stable from the collection time through shipping to the clinical testing site. The stability of the samples at the stated storage conditions will need to be demonstrated to achieve successful validation.

7.4.3.5 *Analyte Concentration Specifications*

Upon establishment of assay performance characteristics relative to a synthetic target in assay qualification, further study during validation is required to understand the potential and ideal range of input from a true clinical sample. First, a significant history of established data on the expected yield of the analyte helps lay the groundwork and ensure that sufficient material will exist to run the device. This range of expected values can then be expanded upon to push the limits of the assay and applied to the studies required in this section. Studies encompassed by this portion of validation include establishment of upper limit of detection and quantitation, lower limit of detection and quantitation, linearity, and range. Data gained from these studies will help determine if additional clinical sample concentration or clinical sample dilution methods will be necessary for the sample prior to analysis. To further address the earlier ease of use section, ideally the assay will accommodate a range of expected concentrations and yields to avoid additional dilution or concentration of the samples.

7.4.3.6 *Cut-off Establishment*

The establishment of an assay cut-point is a key element of the device and determines how robust the performance will be. The identification of a strong cut-point utilizing the training set to assign the threshold and test set, to evaluate the accuracy, has been described previously in this chapter, and thresholds that clearly partition the scores from the patient populations allow the determination of a robust predictive value of the device. Positive or negative shifts to the cut-point will shift the predictive value in either positive or negative favor, which can modify the preferred patient profile. For example, the profile of patients targeted from an assay that has a high negative predictive value based on concordance between low test scores and negative clinical outcome and a more modest positive predictive value based on high test scores and a positive clinical outcome can be drastically modified if the cut-point is increased by a certain factor. The negative predictive value could become suboptimal to the positive predictive value, thus altering the intended implementation and interpretation of the assay.

Patient scores that are more uniformly distributed, or present no level of multimodality, can be more challenging to identify a cut-point in, even with change-point regression models or unimodality tests [23]. It is important to note that most supervised classification methods can *appear* to construct a dividing plane between any two populations with minimal or no errors on a training set of data. However, if this dividing plane or cut-point does not repeat on an independent set of patients, it is not a viable decision rule or threshold. Assays with indeterminate cut-points, or *gray areas*, complicate the interpretation of results by physicians and careful consideration should be given to the value of the assay when discrimination is not apparent. Various methods have been developed to account for such instances, where the values close to the cut-point, or *gray zone* values, are regarded as less confident prediction scores and, as a result, are either not reported or down-weighted in the final assay test result [24,25]. The approach proposed by Coste et al., uses the training set to establish post-test probability scores that are used to calculate likelihood ratio positive and negative thresholds, or upper and lower acceptance values that flank the cut-point at predefined predictive values in setting a *gray zone* [25]. When prediction scores fall in this *gray zone*, the scores are regarded as low confidence scores and not reported. The obvious concern with such a method is the absence of test results for scores that fall in this region. These non-reported results can be combined with assay failure frequencies since, in both cases, no test result is provided to the patient, thus questioning the predictive value of the assay.

The studies included in this section should demonstrate data from the training and test sets and the potential predictive value demonstrated in these two data sets. In addition, the distribution of the signature positive and negative subjects, and where this cut-point lies, should be apparent.

7.4.3.7 *Controls and Calibrators*

The use of assay controls allows for the assessment of fluctuations in the assay and the establishment of run pass/fail criteria. While the identification and use of controls and calibrators may seem trivial in the research setting, it becomes increasingly meaningful as the assay is developed and validated. In a research setting, a single sample 'normal' or a pooled 'normal' created from a number of donors can often be used as a calibrator. However, the practical applications of such a sample become much more difficult under expanded testing in a clinical setting. It becomes increasingly difficult to produce a consistent calibrator sample in this fashion given the constraints of the testing environment. For this reason it is important to identify and implement appropriate controls early in assay development, and limit the need for calibrators if possible. Controls and calibrators can also be generated from a synthetic target, and this should be considered as a viable option as the production of large-scale, highly consistent synthetic targets is logistically more feasible than implementing subject sample controls. Upon identification and implementation of a control and calibration scheme, data included in validation from this section should demonstrate the reproducibility of the controls and calibrators, including the setting of pass and fail criteria based on this performance. In addition, fluctuations in calibrator performance and their effect on assay results should be clearly demonstrated.

7.4.3.8 Precision

The reliability and reproducibility of any assay must be clearly demonstrated, and is the focus of assay validation. The first step in defining the precision of the device is to identify the key sources of variability. For some assays this is an easy process, yet for others it is much more complex. Achieving precision of the device should be performed well in advance of the actual analytical validation. Precision studies will then be designed for each of the key sources of variability. For an example, we will utilize a hypothetical assay whose primary sources of variability are the operator, day of run, and instrument. Typically factorial designs are utilized to test the contribution of each of the sources of variability. To assess the contribution of the sources to the variation, the samples are kept consistent throughout. The following is an example of a factorial design for the theoretical assay being utilized (Fig. 7.7).

This same simple design can be expanded to address additional sources of variability as necessary. It is important to note that as additional factors are added to this design, the number of tests increases on an exponential scale; therefore, only the key sources of variability should be included. Statistical models such as variance components or mixed effects can then be used to identify and quantify the contribution of each factor to the observed variability. As these studies are the focus of analytical validation, properly powering them and evaluating samples that are representative of the intended patient population are critical.

7.4.3.9 Analytical Specificity

Each assay comprising the companion diagnostic device should be specific to its particular analyte. Cross reactivity of the analytes with other intended targets can significantly compromise the reliability of the assay call. Consideration for the cross reactivity of each assay should be investigated early in the development of the assay. For a nucleic acid-based assay, basic logical alignment search tool (BLAST) searches and other sequence alignment methods to assess the overlap of the target with similar targets are useful tools in this effort. Careful attention to the design of assays in the development phase will save significant time and ease the assessment of specificity. Specificity should be demonstrated through specific cross reactivity studies utilizing synthetic targets. Each assay should be tested for its cross reactivity with other assays in the device to ensure it is specifically capturing data for the target of interest and not giving signal from other targets. A graphical representation of this type of cross reactivity is shown below with a theoretical device consisting of five assay targets (Fig. 7.8).

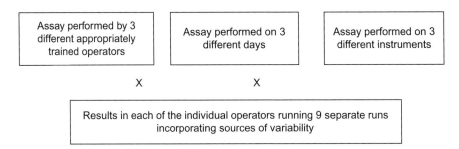

FIGURE 7.7   A graphical representation of a standard factorial design utilized in assay validation to test multiple sources of variation.

Another aspect of cross reactivity is interfering substances. Interfering substances refers to factors present in the sample that may inhibit or in other ways impact a test result. In the case of subjects in a trial or patients in a clinical setting, an example could be some medication that they are currently on or it could be something in the sample preservation. A few examples of interfering substances would be paraffin in a formalin-fixed paraffin-embedded (FFPE) tissue, or the presence of steroid in an inflammatory disease patient. Substances such as these will need to be artificially spiked in to assess the impact of these substances on the assay performance as well as to demonstrate they do not affect the ultimate device call.

7.4.3.10 Assay Conditions

The specific conditions under which the assay is performed should also be defined and specified as part of validation. This is a separate set of specifications from those defined for the sample and refers to the cycling conditions, instrument read settings etc. utilized to generate results for the companion diagnostic assay. In addition to stating the conditions at which the assay is to operate, there should be instrument calibration and monitoring to ensure compliance with the stated assay conditions. Much of this information is included in the clearance of the device, but should be demonstrated for each assay applied to that particular platform.

7.4.3.11 Sample Carryover

The possibility of sample must be addressed. Sample carryover refers not only to the carryover of sample from assay wells, but also to the possibility of the instrument detection carryover from well to well. Studies must be designed with specific markers so that any carryover can be detected by assay specificity. For sample carryover from well to well, instrument calibration and certification information should contain information

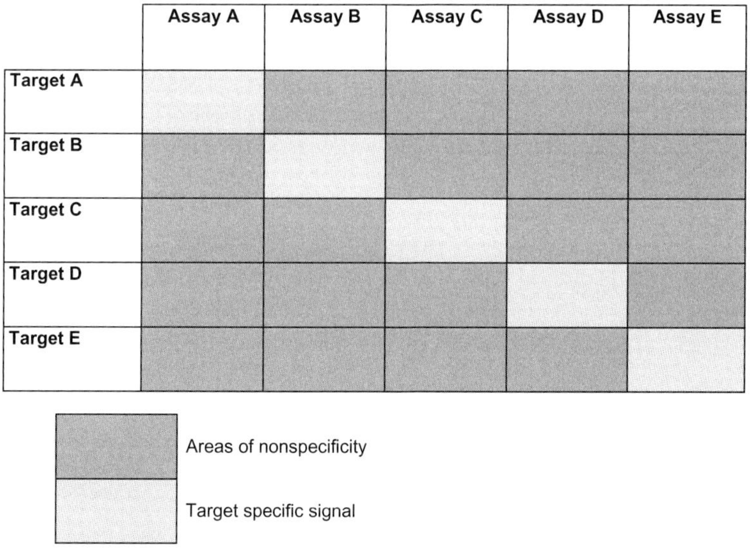

FIGURE 7.8 A diagram depicting a standard cross reactivity study, by which each assay is tested against each target to ensure no cross reactivity of one assay with any of the other analytes.

pertaining to signal leak between wells. Studies must be designed to demonstrate a lack of bleed over of signal from well to well if previous instrument validation parameters do not address this.

7.4.3.12 *Limiting Factors*

The limiting factors category is a general summary of the performance weaknesses of the companion diagnostic assay identified throughout the discovery and validation of the biomarker and assay. Limiting factor analysis identifies key components of the assay and how variations in these components affect potential impact. These issues should be identified, and appropriate mitigation strategies provided for each risk. Comprehensive review of these limiting factors and recognition of appropriate mitigation strategies will be necessary to achieve a full validation.

7.4.4 Clinical Validation

The clinical validation of the companion diagnostic device is the process by which the predictive ability of the device is assessed in the context of a pivotal therapeutic product trial. The topic of how to properly design a pivotal clinical study utilizing a potential companion diagnostic device is a separate topic that is not covered in this chapter. However, ideally the analytically validated market-ready device should be utilized in the enrollment of the pivotal trial to ensure confidence in the call and its predictive ability.

7.5 CONCLUSIONS

Although the process of migrating an assay from a research setting through validation and ultimately clinical implementation can be quite cumbersome, having an awareness of time-sensitive deliverables, and adherence to the guidance provided by the FDA can greatly mitigate potential bottlenecks. Understanding the role of the regulatory authority is critical to the development process, as this is the ultimate device filing decision enabling the use of an assay in a clinical testing environment as a companion diagnostic. The topics presented in this chapter have provided the framework for navigating this process; however, it is a case-by-case scenario to successfully apply the process to each assay on an individual basis, primarily due to the variability in the technical nature of assays. As the feedback from the regulatory agency is crucial to the ultimate approval of the test, it will be important to continually monitor the rationale of the appropriate regulatory authority. While the current methods being utilized in the diagnostic space can be molded to fit this framework, as technical methods such as deep sequencing are applied in the diagnostics space, there will likely be an evolution in analytical processes to accommodate these technical advances.

References

[1] US Department of Health and Human Services, Food and Drug Administration, Center for Devices and Radiological Health, Office of *In vitro* Diagnostic Device Evaluation and Safety, Center for Biological Evaluation and Research. 'Draft Guidance for Industry and Food and Drug Administration Staff: *In vitro* Companion Diagnostic Devices.' Retrieved from <http://www.fda.gov/MedicalDevices/DeviceRegulationandGuidance/GuidanceDocuments/ucm262292.htm>; 2011 [accessed December 2012].

[2] US Department of Health and Human Services, Food and Drug Administration, Center for Devices and Radiological Health, Office of *In vitro* Diagnostic Device Evaluation and Safety, Center for Biological Evaluation and Research. 'Draft Guidance for Industry, Clinical Laboratories, and FDA Staff: *In vitro* Diagnostic Multivariate Index Assays.' Retrieved from <http://www.fda.gov/MedicalDevices/DeviceRegulationandGuidance/Guidance Documents/ucm079148.htm>; 2007 [accessed December 2012].

[3] Department of Health and Human Services, Food and Drug Administration. 'Drug Diagnostic Co-Development Concept Paper.' Retrieved from <http://www.fda.gov/downloads/Drugs/ScienceResearch/ResearchAreas/ Pharmacogenetics/UCM116689.pdf>; 2005 [accessed December 2012].

[4] Lievre A, Bache JB, Le Corre D, Boige V, Landi B, Emile JF, et al. KRAS mutation status is predictive of response to cetuximab therapy in colorectal cancer. Cancer Res 2006;66:3992–5.

[5] Pollack A. Medical treatment, out of reach. The New York Times. Retrieved from <http://www.nytimes. com/2011/02/10/business/10device.html?_r=0&adxnnl=1&pagewanted=all&adxnnlx=1353931534-ltQxlle3jOs- nzMKz1Nu1fQ>; 2011 [accessed December 2012].

[6] Spencer J, Walsh J. FDA rips Europe's system for medical device reviews. Star Tribune, Retrieved from: <http://www.startribune.com/business/148313295.html?refer=y>; 2012 [accessed December 2012].

[7] Otsuji H, Minister of Health, Labour and Welfare. Ministerial Ordinance on Standards for Manufacturing Control and Quality Control for Medical Devices and *In vitro* Diagnostic Reagents. Retrieved from: <http://www.bsigroup. jp/upload/PS-Assessment+Certification/Documents/mdgmp_qms.pdf>; 2004 [accessed December 2012].

[8] SFDA. Regulations for the Supervision and Administration of Medical Devices. Retrieved from <http://www. emergogroup.com/files/china-medical-device-regulations.pdf>; 2000 [accessed December 2012].

[9] European Commission, Enterprise and Industry. Directive 98/79/EC of the European Parliament and of the Council of 27 October 1998 on in vitro diagnostic medical devices. Retrieved from: <http://eur-lex.europa.eu/ LexUriServ/LexUriServ.do?uri=CELEX:31998L0079:en:NOT>; 2008 [accessed December 2012].

[10] Center for Devices and Radiologic Health. The New 510(k) Paradigm, Alternative Approaches to Demonstrating Substantial Equivalence in Premarket Notification: Final Guidance. Retrieved from: <http://www.fda.gov/ MedicalDevices/DeviceRegulationandGuidance/GuidanceDocuments/ucm080187.htm>; 1998 [accessed December 2012].

[11] US Food and Drug Administration. Device Approvals and Clearances. Retrieved from: <http://www.fda.gov/ MedicalDevices/ProductsandMedicalProcedures/DeviceApprovalsandClearances/default.htm>; 2012 [accessed December 2012].

[12] Hastie T, Tibshirani R, Friedman JH. The elements of statistical learning. New York: Springer-Verlag; 2001.

[13] McLachlan GJ. Discriminant analysis and statistical pattern recognition. New York: Wiley; 1992.

[14] Ambroise C, McLachlan G. Selection bias in gene extraction on the basis of microarray gene expression data. Proc Natl Acad Sci 2002;99:6562–6.

[15] Bousquet O, Elisseeff A. Stability and generalization. J Mach Learn Res 2002;2:499–526.

[16] Cawley GC, Talbot NC. On over-fitting in model selection and subsequent selection bias in performance evaluation. J Mach Learn Res 2010;11:2079–107.

[17] Malone B. Big changes coming to molecular Dx reimbursement. Clin Lab News 2011;37(9) Retrieved from: <http://www.aacc.org/publications/cln/2011/September/Pages/MolecularDxReimbursement.aspx#>.

[18] Bellman RE. Dynamic programming. Princeton University Press; 1957. ISBN 978–0–691–07951–6.

[19] Bellman RE. Adaptive control processes: a guided tour. Princeton University Press; 1961.

[20] Donoho D. High-dimensional data analysis: the curses and blessings of dimensionality. Aide-Memoire of a Lecture at AMS conference on Math Challenges of twenty-first Century. 2000.

[21] Kohavi R, John GH. Wrappers for feature subset selection. Artif Intell 1996;97:273–324.

[22] US Department of Health and Human Services, Food and Drug Administration, Center for Devices and Radiological Health, Office of *In vitro* Diagnostic Device Evaluation and Safety, Center for Biological Evaluation and Research. Guidance for Industry, Analytical Procedures and Methods Validation; Chemistry, Manufacturing, and Controls Documentation Draft Guidance. Retrieved from: <http://www.fda.gov/downloads/Drugs/ GuidanceComplianceRegulatoryInformation/Guidances/UCM122858.pdf>; 2000 [accessed December 2012].

[23] Hartigan JA, Hartigan PM. The dip test of unimodality. Ann Statist 1985;13(1):70–84.

[24] Giroti RI, Verma S, Singh K, Malik R, Talwar I. A grey zone approach for evaluation of 15 short tandem repeat loci in sibship analysis: a pilot study in Indian subjects. J Forensic Leg Med 2006;14(5):261–5.

[25] Coste J, Jourdain P, Pouchot J. A gray zone assigned to inconclusive results of quantitative diagnostic tests: application to the use of brain natriuretic peptide for diagnosis of heart failure in acute dyspneic patients. Clin Chem 2006;52(12):2229–35.

Index

Note: Page numbers followed by *"f"* *"t"* and *"b"* refers to figures, tables and boxes respectively.